BRITISH NEMERTEANS

Synopses of the British Fauna

The *Synopses of the British Fauna* are illustrated field and laboratory pocket-books designed to meet the needs of amateur and professional naturalists from sixth-form level upwards. Each volume presents a more detailed account of a group of animals than is found in most field-guides and bridges the gap between the popular guide and more specialist monographs and treatises. Technical terms are kept to a minimum and the books are therefore intelligible to readers with no previous knowledge of the group concerned.

Volumes 1–18 inclusive are published by Academic Press

1. *British Ascidians* R. H. Millar
2. *British Prosobranchs* Alastair Graham
3. *British Marine Isopods* E. Naylor
4. *British Harvestmen* J. H. P. Sankey and T. H. Savory
5. *British Sea Spiders* P. E. King
6. *British Land Snails* R. A. D. Cameron and Margaret Redfern
7. *British Cumaceans* N. S. Jones
8. *British Opisthobranch Molluscs* T. E. Thompson and Gregory H. Brown
9. *British Tardigrades* C. I. Morgan and P. E. King
10. *British Anascan Bryozoans* J. S. Ryland and P. J. Hayward
11. *British Freshwater Bivalve Mollusca* A. E. Ellis
12. *British Sipunculans* P. E. Gibbs
13. *British and Other Phoronids* C. C. Emig
14. *British Ascophoran Bryozoans* P. J. Hayward and J. S. Ryland
15. *British Coastal Shrimps and Prawns* G. Smaldon
16. *British Nearshore Foraminiferids* John W. Murray
17. *British Brachiopods* C. H. C. Brunton and G. B. Curry
18. *British Anthozoans* R. L. Manuel
19. *British Planarians* I. R. Ball and T. B. Reynoldson
20. *British Pelagic Tunicates* J. H. Fraser
21. *British and Other Marine and Estuarine Oligochaetes* R. O. Brinkhurst
22. *British and Other Freshwater Ciliated Protozoa: Part I* C. R. Curds
23. *British and Other Freshwater Ciliated Protozoa: Part II* C. R. Curds, M. A. Gates and D. McL. Roberts

(*Further titles are in preparation*)

A NEW SERIES

Synopses of the British Fauna

No. 24

Edited by Doris M. Kermack and R. S. K. Barnes

BRITISH NEMERTEANS

Keys and notes for the identification of the species

RAY GIBSON

Department of Biology,
Liverpool Polytechnic, Byrom Street,
Liverpool, England

1982

Published for

The Linnean Society of London

and

The Estuarine and Brackish-water Sciences Association

by

Cambridge University Press

Cambridge

London New York New Rochelle

Melbourne Sydney

Published by the Press Syndicate of the University of Cambridge
The Pitt Building, Trumpington Street, Cambridge CB2 1RP
32 East 57th Street, New York, NY 10022, USA
296 Beaconsfield Parade, Middle Park, Melbourne 3206, Australia

First published 1982

Printed in Great Britain at the Pitman Press, Bath

Library of Congress catalogue card number: 81-18193

British Library Cataloguing in Publication Data

Gibson, Ray
British nemerteans.—(Synopses of the British fauna. New series; no. 24)
1. Nemertinea—Great Britain
I. Title II. Linnean Society of London
III. Estuarine and Brackish-water Sciences Association IV. Series
595.1'24 QL391.N6

ISBN 0 521 24619 9 hard covers
ISBN 0 521 28837 1 paperback

P.P.

A Synopsis of the British Nemerteans

RAY GIBSON

Department of Biology, Liverpool Polytechnic,
Byrom Street, Liverpool

Contents

Foreword

Synopsis 24 is a landmark in the study of the British nemerteans (= nemertines). Hitherto their identification has been by comparing specimens with the illustrations in Part 1 of W. C. McIntosh's Ray Society monograph on the British annelids published more than a hundred years ago. This rare and beautiful work is hardly suitable for field and laboratory use, nor is it adequate scientifically, which make a modern key and monograph very welcome. Despite the work of the author, the British nemertean fauna is still comparatively poorly understood, with many species still known only from inadequate original descriptions. This *Synopsis* should make more precise identification possible and it is hoped that more material of these poorly known species will now be recognised for what it is.

Nemerteans are by no means uncommon animals and several are very long; for example, *Lineus longissimus*, the boot-lace worm, is known to attain the length of a Blue whale (30 m) and may exceed twice this. They are also not without scientific importance; witness their role in Bergh's and Goodrich's gonocoel theory for the origin of the coelom. There has even been an attempt to see in them the ancestor of the vertebrates (Willmer, 1974). Many nemerteans are extremely beautiful and Ray Gibson has succeeded in portraying their charm in his illustrations and their fascination in his text. We hope that he will convert many to the study of these 'flatworms' with proboscides, blood vessels and through-guts.

Doris M. Kermack
The Linnean Society

R. S. K. Barnes
Estuarine and Brackish-water
Sciences Association

Introduction

> At all events, whether we are intruding or not, in turning this stone, we must pay a fine for having done so; for there lies an animal as foul and monstrous to the eye as 'hydra, gorgon, or chimaera dire' . . . Its name, if you wish for it, is Nemertes. (Kingsley, 1859)

> The animal, like all its allies . . . is nevertheless exceedingly beautiful. Readers of Kingsley's *Glaucus* will perhaps protest at the adjective, remembering the pages of energetic vituperation which the author hurls at the unfortunate animal, but I cannot think that anyone who studies it without prejudice can fail to be struck by the beauty of the animal. (Newbigin, 1901)

The earliest reference to a nemertean is the brief account of a 'Sea Long-Worm' given by William Borlase in 1758. The phylum now contains an estimated 900 species (Gibson, 1982), many of which are poorly known and inadequately described. Nemerteans are typically soft-bodied, free-living, benthic marine worms found intertidally or sublittorally burrowed into soft muddy or sandy sediments, creeping amongst algal fronds and holdfasts or over colonial sessile invertebrates, in crevices, or beneath boulders and rocks. Some species live on land or in freshwater, a few are commensal or parasitic, and one group is entirely pelagic.

McIntosh (1873–74), in the last text devoted exclusively to the British nemerteans, listed 31 species, all of which were marine. Although the taxonomic status of several remains uncertain, there are now 74 species of Nemertea (about 8% of the world total) recorded from the British Isles, including one from terrestrial and two from freshwater habitats.

Nemerteans, indeed, are a poorly investigated invertebrate group, principally because of the difficulties involved in identifying them. The problem is that their taxonomy and identification are primarily based upon internal morphology and this can only be adequately studied by the use of laborious histological procedures. Often, too, the fragile nature of their bodies means that only fragmented or incomplete specimens are collected (this is especially true of nemerteans obtained from dredge or grab samples), such that taxonomically important parts of the body (e.g. the cephalic region or the proboscis) are lost. With these difficulties to contend with it is hardly surprising that most zoologists have either preferred to ignore nemerteans entirely or only to identify them as members of this phylum. Even when generic or specific names are provided the identifications are often based only upon external features and relatively few nemerteans can be reliably identified without reference to their internal anatomy. Indeed, comparatively few species have either been adequately described or had any aspect of their biology investigated in detail. Accordingly, the species descriptions provided here must not be regarded as complete, nor should the generic placing of species or the arrangement of the families be taken as necessarily correct, since many nemertean species, genera and families are urgently in need of taxonomic revision. At present, however, we lack sufficient data on most taxa for useful revisions to be achieved.

Users of this book, in depending for their identifications upon a key which does not primarily involve the use of histological techniques, will inevitably find specimens which do not adequately key out. The author, or the Linnean Society who will pass them on, would welcome such material for examination or, indeed, any information concerning the British nemerteans.

General morphology

The account of nemertean morphology given here specifically relates to the British fauna. For additional information the reader should consult Gibson (1972).

External features

Nemerteans are characteristically elongate, vermiform animals with soft bodies, often capable of extreme contraction and elongation, which range in length from a few millimetres (*Carcinonemertes*, *Tetrastemma*) up to 30 m (*Lineus*); most, however, are less than about 20–30 cm long. The length, width and shape of the body depend upon the degree of contraction or extension but many benthic species are less than one millimetre wide and elliptical or rounded in section (Fig. 1A). Others, especially those with a dorsoventrally flattened shape (Fig. 1B), may be 6–8 mm or more wide in the intestinal regions and occasional individuals, such as a large *Cerebratulus marginatus*, may have a width of about 25 mm. *Malacobdella grossa*, entocommensal in bivalve molluscs, possesses a leech-like shape (Fig. 1C); members of this genus are the only nemerteans with a **posterior ventral sucker**.

Nemerteans do not possess a distinct head. Anteriorly they are mostly pointed, rounded or blunted, although in some genera (*Amphiporus*, *Tetrastemma*, *Tubulanus*) there is an anterior **cephalic lobe** (Fig. 1D, E) which is lanceolate, spatulate, semi-circular or heart-shaped; this does not strictly constitute a head as it does not normally house the cerebral ganglia. The cephalic region of many species bears transverse, oblique or longitudinal furrows, the **cephalic grooves** or **slits** (Fig. 1F); these are most strikingly developed in the Heteronemertea, especially the Lineidae where they form deep horizontal lateral grooves (Fig. 1G). The anterior region also contains the **eyes**, although many species, at least as adults, do not possess them (*Cephalothrix*, *Malacobdella*). Eye number varies both inter- and intraspecifically, ranging from two (Fig. 1H) to many (Fig. 1I, J). The eyes are arranged more or less bilaterally. In species with two to six eyes the number is usually constant, whereas in other forms eye number often increases with age and size. Frequently the eyes are partially or almost completely obscured by the cephalic pigmentation.

The **proboscis pore** opens at or just below the anterior tip in most nemerteans, but in the heteronemertean family Valenciniidae it is located further back and in some genera lies close to the cerebral ganglia. The mouth

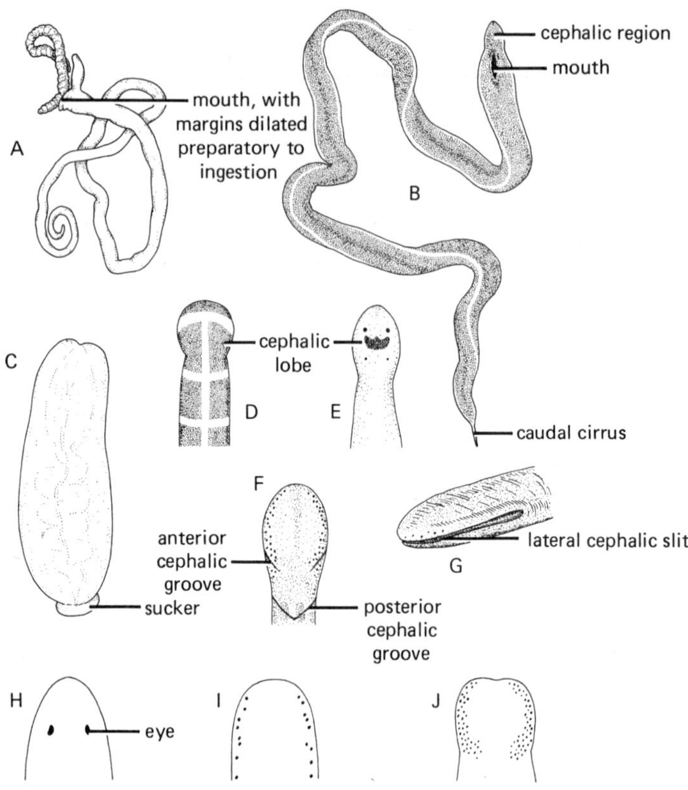

Fig. 1. Diagrams to show some external features of nemerteans. Examples of general body shape are shown in A, *Cephalothrix linearis*; B, *Cerebratulus marginatus*; and C, *Malacobdella grossa*. The cephalic regions (D–J) are drawn to emphasis the appearance of cephalic lobes in D, *Tubulanus annulatus*; E, *Tetrastemma peltatum*; the two pairs of oblique cephalic grooves in F, *Amphiporus dissimulans*; the lateral horizontal cephalic slits of G, *Lineus ruber*; and typical patterns of eye distribution in H, *Carcinonemertes carcinophila*; I, *Lineus lacteus*; and J, *Baseodiscus delineatus*.

and proboscis pore are separate in the Anopla; in the palaeonemertean family Cephalothricidae the mouth forms a small ventral aperture some distance behind the cerebral ganglia at the end of a long slender snout (Fig. 1A), in other Palaeonemertea and in the Heteronemertea it is located below or just behind the brain lobes (Fig. 1B). During feeding the mouth can be enormously dilated to facilitate the ingestion of large prey organisms (Fig. 1A). In the Enopla the mouth and proboscis pore usually open via a common anterior aperture but in some, particularly the benthic Polystilifera (*Paradrepanophorus*, *Punnettia*), they open independently either directly on the cephalic surface or indirectly through an **atrial chamber**. In members of the Enopla with separate oral and proboscis apertures the mouth is always located close to the proboscis pore and anterior to the cerebral ganglia.

The posterior end of the body usually either tapers gradually to a sharp or blunt tip (Fig. 1A) or, occasionally, as in the heteronemertean genera *Cerebratulus*, *Micrella* and *Micrura*, terminates in a slender tail or **caudal cirrus** (Fig. 1B). The **anus** opens at or just dorsal to the posterior end (Fig. 4D, E), or at the base of the caudal cirrus when one is present. In the Bdellonemertea (*Malacobdella*) the anus is terminal but above the sucker (Fig. 1C, 4F).

Many species are more or less uniformly white or cream coloured or tinted in shades of grey, brown, red, orange, pink, green or yellow. In paler coloured forms, or in those in which the body is translucent, internal organs such as the proboscis and rhynchocoel, the gut, the gonads and the cerebral ganglia are often discernible through the body wall (Fig. 2). In some species colour changes apparent during the reproductive period can be attributed to pigments contained within the mature gonads, and differences between the sexes can lead to a colour dimorphism. Colour variations in the intestinal regions may be due to the gut contents. There are many nemerteans, however, which possess striking and distinctive colour patterns. These patterns, which may be composed of stripes, bands, speckles, marbled effects or contrasting colours arranged in geometric shapes, are often species specific and are frequently more clearly marked on the dorsal surface (Fig. 1D) or are confined to the cephalic regions (Fig. 1E).

Internal anatomy

Nemertean worms are unsegmented, bilaterally symmetrical, acoelomate invertebrates characterised by an eversible muscular **proboscis** contained within a dorsal fluid-filled chamber, the **rhynchocoel**, a blood system composed of distinct vessels, and an alimentary tract with separate mouth and anus (Fig. 2). Subdivision of the phylum into its lower taxa is dependent upon internal morphology, particularly the organisation of the body wall, the nervous system and the proboscis apparatus.

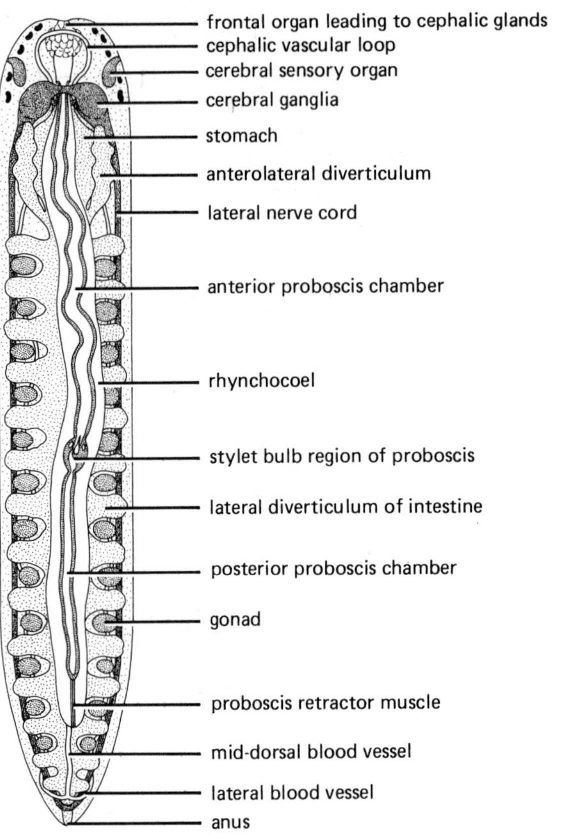

Fig. 2. Schematic diagram to show the arrangement of internal structures of a generalised monostiliferous hoplonemertean in dorsal view.

The body wall (Fig. 3) comprises three principal components, the **epidermis**, the **dermis** and the **body wall musculature**. All nemerteans possess a ciliated epidermis composed of glandular, sensory and interstitial cells, but in *Carinoma* the epidermis of the anterior body is unusual in containing an **intra-epithelial muscle fibre network** (Fig. 9B). Greater differences are found in the arrangement of the dermis. In the Palaeo-, Hoplo- and Bdellonemertea the dermis consists of hyaline connective tissue which, dependent upon the species, varies from a simple epidermal basement membrane to a thick and distinct layer. Within the Heteronemertea, however, the dermis is predominantly fibrous, often very much thicker than the epidermis and characteristically associated with gland cells of various types. The gland cells may be arranged into a distinct peripheral layer (Fig. 3C), separated from the body wall muscles by an inner connective tissue region (*Baseodiscus*, *Euborlasia*, *Poliopsis*) or the glands and fibrous tissue may be intermingled and dispersed between the fibres of the outer muscle layer (*Micrella*, *Oxypolia*, *Valencinia*); both arrangements occur in *Cerebratulus*, *Lineus* and *Micrura*. The sequence of muscle layers in the body wall is characteristic of the various orders. In the Palaeonemertea the number of major layers is either two (outer circular, inner longitudinal: *Carinoma*, *Cephalothrix*, *Procephalothrix*) or three (outer circular, middle longitudinal, inner circular: *Carinesta*, *Tubulanus*) (Fig. 3A, B), although additional and incomplete layers restricted to the anterior body regions occur in *Carinoma* and *Procephalothrix*. Several of the tubulanid species possess dorsal and/or ventral **muscle crosses** between the two circular layers. The Heteronemertea possess three principal muscle layers (Fig. 3C) (outer longitudinal, middle circular, inner longitudinal), but an extra, diagonal, zone is found in many species of *Cerebratulus*. In Hoplo- and Bdellonemertea the body wall musculature basically comprises outer circular and inner longitudinal layers (Fig. 3D), although in some monostiliferous hoplonemerteans (*Amphiporus*, *Nipponnemertes*) there is a middle diagonal zone and in the polystiliferous hoplonemertean genera a residual (*Punnettia*) or well developed (*Paradrepanophorus*) inner circular muscle layer is present.

Other muscle systems of the body may be of taxonomic significance. **Dorsoventral muscles**, typically passing between the intestinal diverticula, are especially strongly developed in species capable of swimming actively (*Cerebratulus*). **Splanchnic muscles** ensheathing the foregut (Fig. 3C, 7B) and variably composed of longitudinal, circular or oblique fibres are characteristic of many heteronemertean genera. In *Cephalothrix* and *Procephalothrix* a distinct **horizontal muscle plate** lies between the foregut and rhynchocoel (Fig. 3A); a similar structure also occurs in many Heteronemertea. The rhynchocoel wall too possesses its own musculature (Fig. 3); in most nemerteans this consists of separate circular and longitudinal layers, but in *Argonemertes*, *Nipponnemertes*, *Oerstedia*, *Paradrepanophorus* and *Punnettia* the muscles form a meshwork of interwoven longitudinal and circular

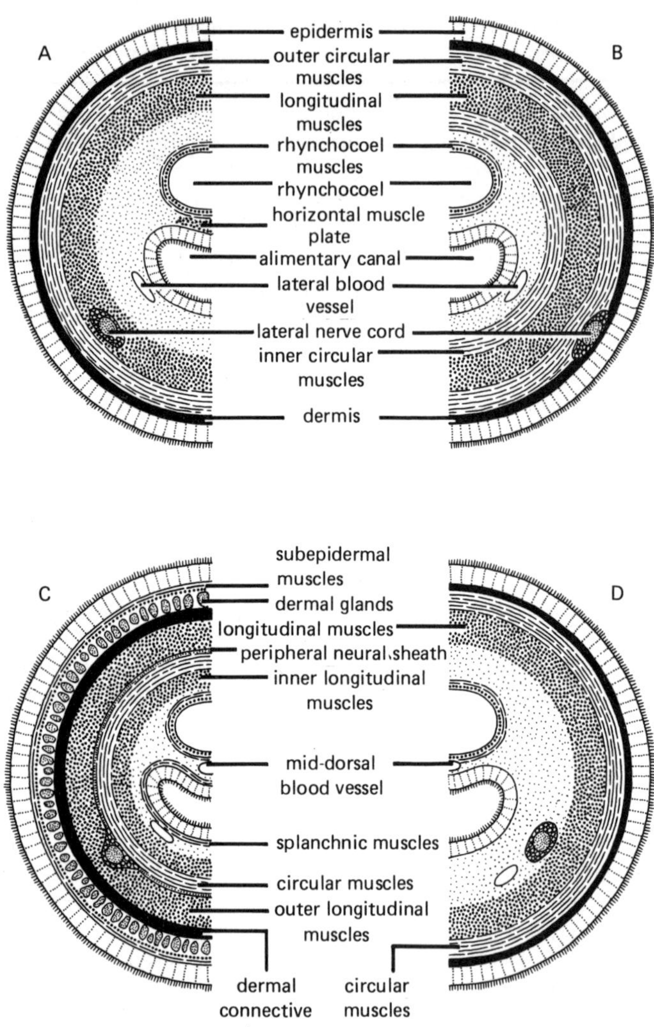

Fig. 3. Schematic diagrams to show the organisation of the body wall and the position of the principal organ systems in the different nemertean orders: A, B, Palaeonemertea; C, Heteronemertea; D, Hoplonemertea and Bdellonemertea. Redrawn from Gibson (1982).

fibres. In *Carcinonemertes* the rhynchocoel wall consists only of a thin lining membrane and lacks muscles.

The nervous system of nemerteans (Fig. 2) consists of a pair of bilobed **cerebral ganglia** transversely connected by **dorsal** and **ventral commissures** (Fig. 4A), a pair of ganglionated longitudinal nerve cords (the **lateral nerves**) and a complex of minor nerves. The position of the principal neural elements relative to the body wall layers is of major taxonomic importance. In the Hoplo- and Bdellonemertea they lie internal to the body wall musculature amongst parenchymatous tissues (Fig. 3D), but in anoplan nemerteans are more peripherally placed. In the Heteronemertea the lateral nerves run between the outer longitudinal and middle circular muscle layers (Fig. 3C), although in *Poliopsis* they extend posteriorly enclosed by the circular fibres. The cerebral ganglia are closely surrounded by the cephalic muscles. The greatest variation occurs amongst the Palaeonemertea; in the Carinomidae the cerebral ganglia and anterior portions of the lateral nerves lie external to the body wall outer circular muscles (Fig. 3B) but in the intestinal region the lateral nerves run amongst the inner longitudinal muscle fibres; in the Cephalothricidae the cerebral ganglia and lateral nerves are located in the body wall longitudinal muscle layer (Fig. 3A); and in the Tubulanidae they are situated between the epidermal basement membrane and the outer circular muscle layer (Fig. 3B).

At the generic level other features of the nervous system are sometimes significant. Each lobe of the cerebral ganglia contains a **central fibrous region** surrounded by a layer of **ganglionic cells** (Fig. 4A). Usually the two regions are separated by a thin but distinct connective tissue stratum, the **inner neurilemma**, with a similar **outer neurilemma** investing the ganglionic lobes as a whole (Fig. 4A). In *Micrura*, however, the outer neurilemma is often missing. Amongst the Heteronemertea the shape of the dorsal fibre core is commonly employed as a generic character; in all the British hetero-nemerteans this core is anteriorly simple but posteriorly forked into upper and lower branches. The lateral nerves, which extend posteriorly from the rear of the ventral cerebral lobes, also contain a fibrous core ensheathed by ganglionic cells, but in some hoplonemertean genera (*Argonemertes*, *Oerstedia*) an **accessory lateral nerve** arises from the dorsal cerebral lobes and runs in the upper portion of each lateral nerve (Fig. 4B, C). **Neurochords** and **neurochord cells** form a characteristic feature of many nemerteans. The neurochord cells are particularly large-sized cells situated in the ganglionic layer of the cerebral lobes (Fig. 4A); there may be one or more pairs. Neurochordal nerve processes extend from these cells into the lateral nerves, where their large size usually makes them very obvious in the nerve fibre tract (Fig. 4C). When numbers of neurochords are present they tend to be aggregated into bundles.

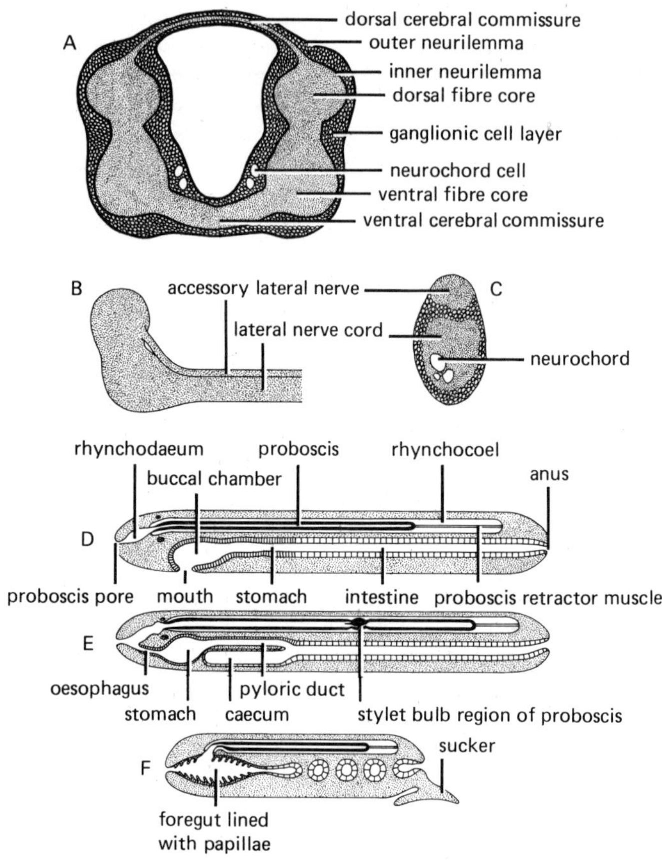

Fig. 4. Diagrams to show aspects of the internal anatomy: A, transverse section through a pair of cerebral ganglia, drawn to emphasise the major anatomical features associated with the central nervous system; B, lateral view of the anterior portions of a lateral and accessory lateral nerve to show their origin from the ventral and dorsal cerebral lobes respectively; C, a lateral and accessory lateral nerve in transverse section; D–F, schematic vertical longitudinal sections to show the relationships of the alimentary tract and proboscis apparatus within the different nemertean orders; D, Palaeonemertea and Heteronemertea; E, Hoplonemertea; F, Bdellonemertea. D–F redrawn from Gibson (1972).

The **proboscis apparatus** includes the **proboscis, rhychodaeum** and **rhynchocoel** (Fig. 4D, E, F). The rhynchodaeum is a tubular chamber opening at the proboscis pore, found in all groups except the Bdellonemertea (Fig. 4F). In the Anopla (Fig. 4D) and in those hoplonemerteans in which the proboscis pore and mouth are separate the rhynchodaeum is quite distinct from the gut, but in most Hoplonemertea the alimentary canal opens from the floor of the rhynchodaeum (Fig. 4E). The junction between rhynchodaeum and rhynchocoel forms the proboscis insertion point, usually located immediately in front of the cerebral ganglia but just behind them in *Carinesta*. The rhynchocoel extends posteriorly above the gut (Fig. 4D, E, F); in most genera it reaches for most or all of the body length, but in *Baseodiscus*, *Emplectonema* and *Nemertopsis* is less than half as long as the body and in *Carcinonemertes* is reduced to a small chamber which barely extends beyond the cerebral ganglia. Lateral **rhynchocoelic diverticula** (Fig. 5A) are present in some genera; they are shallow and restricted to the foregut region in the heteronemertean *Micrella*, but in the polystiliferous Hoplonemertea (*Paradrepanophorus*, *Punnettia*) are long and extend throughout the rhynchocoel length. The nemertean proboscis is a long muscular organ, formed by an invagination of the anterior end of the body. There are basically two distinct types, generally referred to as the unarmed form, characteristic of the Palaeo- and Heteronemertea, and the armed form, representative of the Hoplonemertea. In the Bdellonemertea the proboscis is unarmed but because of its similarities to that of hoplonemerteans is regarded as derived from this type. The unarmed proboscis in some genera (*Cephalothrix*, *Lineus*) is provided with large numbers of **epithelial barbs** (Fig. 5B), but these are not confined to a specific part of the organ as is the armature of the hoplonemerteans. In the British Palaeonemertea there are typically two **proboscis muscle layers** (outer circular, inner longitudinal) (Fig. 5C). Some Heteronemertea (*Euborlasia*, *Lineus*, *Micrella*, *Micrura*) possess the same muscle arrangement, in others (*Baseodiscus*, *Cerebratulus*, *Poliopsis*, *Valencinia*) the sequence is reversed. In several genera (*Cerebratulus*, *Lineus*, *Micrura*, *Oxypolia*) there are representatives with three proboscis muscle layers (outer longitudinal, middle circular, inner longitudinal) (Fig. 5D). **Muscle crosses** in the proboscis wall (Fig. 5D) are a characteristic feature of several heteronemertean genera, there being either one or two crosses depending upon the species. In most instances the unarmed proboscis possesses a uniform construction throughout its length; in some species, however, particularly of *Cerebratulus*, there is a reduction in the number of muscle layers towards the proboscis insertion. In the Hoplonemertea the proboscis is armed with either a single **central stylet** carried on a cylindrical **basis** (Fig. 5E) (Monostilifera) or with a pad- or shield-like basis bearing large numbers of minute stylets (Fig. 5F) (Polystilifera). The monostiliferous proboscis is divisible into three distinct regions, an anterior thick-walled tube, a short median muscular **stylet bulb region** (housing the proboscis armature) and a posterior blind-ending tube. There are typically two

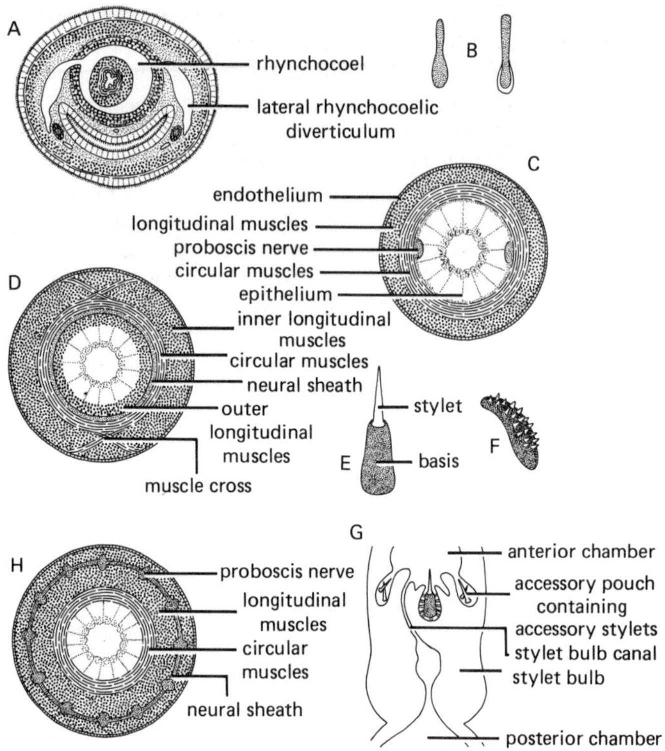

Fig. 5. Diagrams to show the anatomy of the proboscis apparatus: A, transverse section through the body of a polystiliferous hoplonemertean to show a pair of lateral rhynchocoelic diverticula; B, proboscis epithelial barbs of *Cephalothrix linearis* (left) and *Cephalothrix rufifrons* (right); C, transverse section to show the structure of a typical palaeonemertean proboscis; D, transverse section to show the structure of a typical heteronemertean proboscis; E, the central stylet and basis of a monostiliferous hoplonemertean; F, the stylets and basis of a polystiliferous hoplonemertean; G, schematic longitudinal section through the stylet bulb region of a monostiliferous hoplonemertean to show the position of the central stylet and a pair of accessory stylet pouches; H, transverse section to show the structure of a typical hoplonemertean proboscis. B redrawn from Jennings & Gibson (1969).

accessory pouches containing one or more **accessory stylets** in the middle bulb region (Fig. 5G), although *Tetrastemma quatrefagesi* possesses four and in species of *Carcinonemertes* a general reduction in proboscis development has led to the loss of the pouches and their accessory stylets. The proboscis musculature generally comprises inner and outer circular layers enclosing a very much thicker longitudinal zone which is commonly separated into two regions by the well developed proboscis nerve layer (Fig. 5H). The outer circular muscles are often missing from the posterior portion. The polystiliferous proboscis possesses a similar general construction but lacks a distinct muscular stylet bulb. In bdellonemerteans (*Malacobdella*) the proboscis, much simpler in construction than that of other enoplans, opens into the dorsal wall of the spacious foregut and there is no external proboscis pore as such (Fig. 4F)

Other organ systems of the body possess varying degrees of taxonomic importance. The **blood system** exhibits several grades of complexity; the simplest condition, in which a pair of **lateral longitudinal vessels** is transversely connected only by **cephalic** and **anal lacunae**, occurs in the Cephalothricidae (Fig. 6A). In other Palaeonemertea the basic pattern is modified by subdivision of the cephalic lacuna (*Carinesta*), by the development of a subdivided **foregut lacuna** (*Tubulanus*), or by the formation of additional **post-cerebral vessels** in the foregut region (*Carinoma*). None of the British palaeonemerteans possesses a **mid-dorsal vessel**. In the Heteronemertea and the Enopla there are three principal longitudinal vessels (paired lateral and single mid-dorsal) (Fig. 6B), but in many heteronemertean genera additional vessels and lacunae are developed in association with the cerebral and foregut regions. The three main vessels are often linked in the intestinal region by pseudometamerically arranged **transverse connectives** (Fig. 6C) alternating with the **gut diverticula**. These connectives are, however, missing from many Hoplonemertea and the Bdellonemertea; in *Carcinonemertes* the mid-dorsal vessel too is absent. The mid-dorsal vessel usually penetrates the rhynchocoel wall to form either an elongate and often distensible **rhynchocoelic villus** (Fig. 6D) (Heteronemertea) or a distinct rounded **vascular plug** (Fig. 6E) (Hoplonemertea). In *Argonemertes* and *Oerstedia* the mid-dorsal vessel divides for a short distance and each branch bears a plug (Fig. 6F). The blood system in the terrestrial genus *Argonemertes* has evolved an extensive **submuscular capillary network** in addition to the main longitudinal vessels.

A **nephridial excretory system** is present in most nemerteans, typically consisting of one or more pairs of branched tubules located in the parenchyma between the cerebral ganglia and the rear of the foregut (Fig. 7A) and opening to the exterior via a single pair or a few **nephridiopores**. The excretory tubules, particularly in the Heteronemertea, are often in intimate connection with branches of the blood system. No evidence of an excretory system has been found in *Carcinonemertes*, however, and in some species of *Baseodiscus* the **efferent tubules** discharge into the foregut. Increased development of the excretory tubules and multiplication of the nephridiopores is found especially in the terrestrial and freshwater species; in *Argonemertes* there is an extensive network of tubules associated with large numbers of mononucleate **flame cells** (Fig. 38D) (similar flame cells also occur in the related genus *Prosorhochmus*) and in *Prostoma* the main pair of collecting tubules extends the full length of the body.

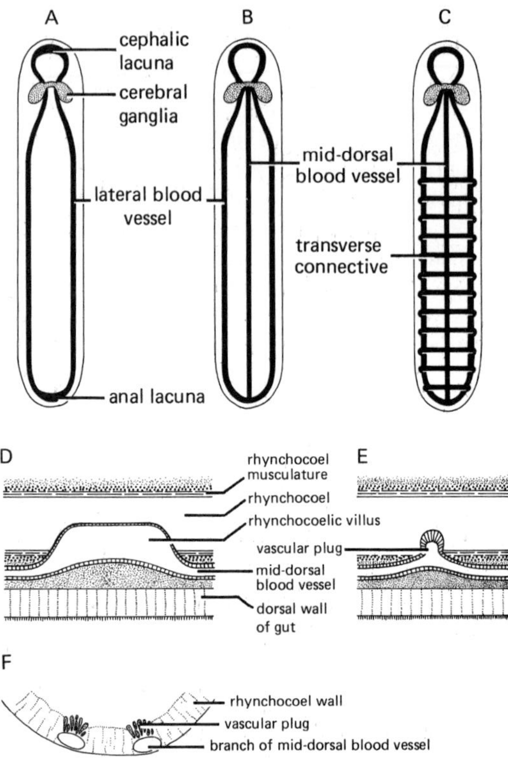

Fig. 6. Diagrams to show the organisation of the blood vascular system. Variations in the degree of complexity of the blood system are represented by the A, cephalothricid palaeonemertean type; B, hoplonemertean type; and C, heteronemertean type. A rhynchocoelic villus, D, and vascular plug, E, are shown in schematic longitudinal view; F, transverse section through the ventral rhynchocoel wall of *Argonemertes dendyi* to show the two vascular plugs of this species. A–C based on Gibson (1972), F redrawn from Gibson (1972) after Pantin (1969).

The **alimentary canal** in all nemerteans consists essentially of a ciliated tube which is divisible both morphologically and functionally into **foregut** and **intestine** (Fig. 4D, E, F). In the Anopla the **mouth** opens from the ventral surface into a glandular **stomach** (Fig. 4D) which, especially in the Heteronemertea, may possess deeply folded walls. There may be a distinct **subepithelial gland-cell zone** ensheathing the stomach, as in *Baseodiscus* (Fig. 7B). The junction between stomach and intestine in anoplans is generally marked only by a change in the nature of the gut wall. In *Poliopsis*, however, the posterior portion of the foregut is attenuated and a short ventral **intestinal**

caecum extends anteriorly below it. The intestine bears deep to shallow pseudometamerically disposed lateral **diverticula** for most of its length (Fig. 2) but these are absent from *Carinesta*. The gut of enoplans is usually more complex. A distinct **oesophagus** (Fig. 4E) is present in many species, connecting the stomach with the mouth or rhynchodaeal chamber. The stomach itself is characteristically bulbous in shape, with its posterior portion developed into an elongate slender duct, the **pylorus** (Fig. 4E). This usually opens into the dorsal wall of the intestine, the anterior part of which extends forward below the pylorus to form an intestinal caecum. A caecum is absent or only poorly developed in *Carcinonemertes*, *Emplectonema* and *Prostoma*. Often the caecum bears a pair of anteriorly directed diverticula (Fig. 2) (missing from *Nemertopsis*, *Nipponnemertes* and *Oerstedia*) and, as in the remaining intestinal regions, lateral diverticula of varying form. The gut in *Malacobdella* is quite different; it consists of a barrel-shaped foregut, lined with longitudinal tracts of motile **papillae**, opening into a sinuous intestine which lacks diverticula (Fig. 4F, 50A, B).

Several types of sense organs, mainly limited to the anterior body regions, occur in nemerteans. Typical of a number of Palaeonemertea are the **epidermal sensory organs**, arranged either as a median dorsal row of ciliated pits on the head (*Carinoma*) (Fig. 7C) or as a pair of lateral organs located in the foregut region of the body (species of *Tubulanus*). Similar structures are also present in the heteronemertean genus *Micrella*, situated laterally just behind the excretory pores. A **frontal organ** (Fig. 2, 7D) is present in most nemerteans. In the Hoplonemertea it is characteristically a single ciliated epidermal chamber opening just above or into the rhynchodaeum (Fig. 41C, D), but in *Prosorhochmus* it is developed into a long duct which penetrates the cephalic tissues. *Argonemertes*, *Carcinonemertes* and some species of *Emplectonema* do not possess a frontal organ. Among the Heteronemertea a few species of *Cerebratulus*, *Lineus* and *Micrura* supposedly lack a frontal organ and none is found in *Micrella* or *Oxypolia*. The organ in this order may be like that of the Hoplonemertea or be replaced by three similar small ciliated pits which either open independently or into a common duct. It is not certain that the frontal organ possesses a sensory function, although in a few instances there is some evidence that it may be chemotactic in nature. Most nemerteans possess mucus-secreting **cephalic glands** (Fig. 2, 7D, E) in their anterior regions; these discharge through the frontal organ when one is present, or via numerous independent ducts opening over the cephalic surface when there is no frontal organ (Fig. 7E). The cephalic glands are mostly restricted to the pre-cerebral parts of the body, but are especially extensive in *Argonemertes*, *Baseodiscus*, *Nemertopsis*, *Oxypolia*, *Prosorhochmus* and *Valencinia* where they extend behind the cerebral ganglia into the foregut regions. Paired **cerebral sensory organs** (Fig. 2, 7F) are a characteristic feature of most nemerteans, although they are absent from many Palaeonemertea (*Carinesta*, *Carinoma*, *Cephalothrix*, *Procephalothrix*) and from the hoplonemertean *Carcinonemertes* and the bdellonemertean

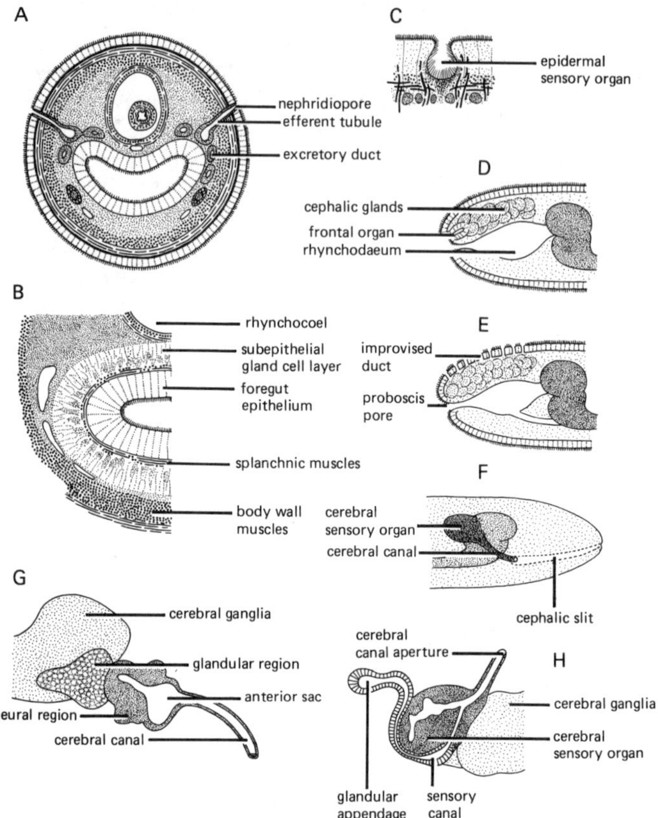

Fig. 7. Diagrams to show aspects of the organisation of the excretory system, foregut, cephalic glands and sensory organs: A, transverse section through the posterior foregut region of a hoplonemertean to show the typical position of the collecting ducts, efferent tubules and nephridiopores of the excretory system; B, transverse section through a part of the foregut region of a baseodiscid heteronemertean to show the well developed subepithelial gland cell layer surrounding the gut; C, longitudinal section through a dorsal epidermal sensory pit of *Carinoma armandi*; D, E, schematic longitudinal sections through the head, showing the cephalic glands either discharging to the exterior through the frontal organ, D, when one is present, or via numerous improvised pores, E, when a frontal organ is missing; F, schematic lateral view to show the position of the cerebral sensory organs and manner whereby they open into the horizontal cephalic slits of a lineid heteronemertean; G, a cerebral sensory organ of *Argonemertes dendyi*; H, a cerebral sensory organ of *Punnettia splendida*. G redrawn from Pantin (1969), H redrawn from Stiasny-Wijnhoff (1934).

Malacobdella. The organs form neuroglandular bodies opening to the exterior via ciliated **cerebral canals** which lead either to separate pores (Fig. 7G) or to the cephalic slits (Fig. 7F). In the Heteronemertea they are always intimately attached to the rear of the dorsal cerebral ganglionic lobes and form large ovoid structures (Fig. 7F); often their posterior surfaces are bathed by blood lacunae. Amongst the Hoplonemertea, however, the cerebral organs exhibit variable degrees of development. They are mostly located in front of the cerebral ganglia (Fig. 2), less often alongside or behind, but are connected with them by distinct nerve tracts. In the terrestrial species *Argonemertes dendyi* the cerebral canal is forked, one branch connecting an **anterior sac** with the body surface (Fig. 7G), the other leading from the sac into the neural part of the cerebral organ. The British Polystilifera belong to the tribe Reptantia (non-pelagic species); this tribe is divided into two groups partly according to the organisation of their cerebral organs. In the Aequifurcata (*Punnettia*) the organs contain two **sensory canals** and a distinct **glandular appendage** (Fig. 7H), whereas in the Inaequifurcata (*Paradrepanophorus*) there is no glandular appendage and only a single sensory canal.

Eyes are found in most of the monostiliferous and non-pelagic polystiliferous hoplonemerteans and in many palaeo- and heteronemerteans. They are typically located amongst the cephalic dermal, muscular or parenchymatous tissues but are externally visible (Fig. 1E, F, H, I, J) unless the body surface is deeply pigmented. The eyes are cup-shaped structures containing a layer of black, brown or reddish pigment granules.

Most species of nemerteans are oviparous and possess separate sexes, although *Prostoma*, *Prosorhochmus* and some *Argonemertes* are hermaphroditic. *Prosorhochmus claparedii* combines hermaphroditism with ovoviviparity and anal parturition, the young worms emerging from the gonads to enter the alimentary tract. The **gonads** of nemerteans are spherical to flask-shaped structures, usually restricted to the intestinal region where they form two lateral rows either more or less serially repeated or alternating singly or in groups with the intestinal diverticula. In the parasitic or commensal forms (*Carcinonemertes*, *Malacobdella*) the gonad number is increased with a concurrent loss in regularity. Each gonad discharges its ripe gametes independently except in male *Carcinonemertes*; in these nemerteans a **vas efferens** links each **testis** with a single **vas deferens**, which in turn leads to a sperm storage chamber (the **seminal vesicle**) from where the gametes are shed into the rear of the intestine and thence through the anus (Fig. 32C). In the ovoviviparous species the ova are retained within the gonads and fertilisation is internal. Usually in anoplan nemerteans several eggs are matured simultaneously within each ovary, whereas in enoplan forms one or only a few eggs are matured at a time.

Biology

Life history

The life-span of nemerteans is unknown. Species of *Lineus* and *Prostoma* have been maintained under laboratory conditions for more than a year, others have survived for much shorter times. Coe (1905) suggests that many species are probably annual, but that several, especially those which attain a large size (such as *Baseodiscus delineatus*, *Cerebratulus marginatus* and *Lineus longissimus*), may live for several years.

Little is known about the natural reproductive habits of nemerteans. Fertilisation is mostly external. In some species (*Lineus ruber*, *Amphiporus lactifloreus*) mature worms associate together in a common gelatinous sheath into which they discharge their eggs and sperm. The fertilised eggs, contained within egg capsules, remain embedded in the sheath matrix after the adults emerge and the egg-string so formed (Fig. 8A) adheres to the surface of rocks or algal fronds. With other nemerteans, such as *Malacobdella*, ripe gametes

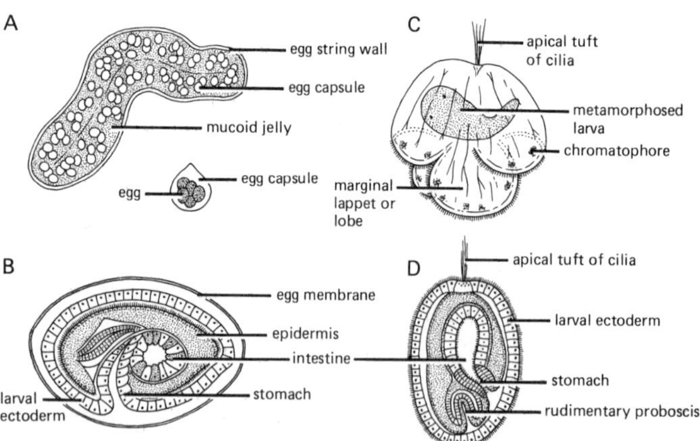

Fig. 8. Diagrams to show a nemertean egg-string and the three types of larvae: A, gelatinous egg-string of *Lineus ruber*, with an enlarged drawing of an individual egg capsule containing five eggs; B, sagittal section through an advanced stage of development of a *Lineus* species Desor larva; note the retention of the egg membrane during development; C, a pilidium larva of a *Cerebratulus* species developed to the stage where it contains a fully metamorphosed but miniature nemertean; D, sagittal section of the Iwata larva of *Micrura akkeshiensis*. All redrawn from Gibson (1972).

are apparently shed directly into the sea and fertilisation occurs without the two sexes coming into direct contact; in such instances the stimulus to discharge the gonad contents is possibly chemical. The terrestrial hoplonemertean *Argonemertes dendyi* deposits gelatinous egg capsules, each containing 8–30 eggs, in wet locations (Waterston & Quick, 1937; Pantin, 1961; Anderson, 1980). So far as is known, most species possess a particular breeding season, even though this may extend over three or four months. *Malacobdella grossa* appears to be the only British species which reproduces throughout the year, although superimposed upon the annual rhythm are peaks in activity which can be correlated with the availability of the phytoplanktonic food of this species (Gibson, 1968a, b).

Two distinct types of embryological development can be recognised in nemerteans. The direct type, found in all orders except the Heteronemertea, involves the embryo hatching from its egg into a larva which grows into a miniature adult without the intervention of a metamorphosal stage. In the Palaeonemertea hatching occurs before the embryonic proboscis has fully differentiated, in the Hoplo- and Bdellonemertea the proboscis is fully formed before the larva emerges from its egg-case. Development in the Heteronemertea, in contrast, involves a distinct intermediate larval stage which undergoes metamorphosis to form an adult. The larval ectoderm, shed during metamorphosis, may be eaten by the emerging juveniles (Cantell, 1966). Three types of larvae are distinguished; the **Desor** larva (Fig. 8B) has no pelagic phase in its development and grows entirely inside the egg membrane, whereas the **pilidium** larva (Fig. 8C) (somewhat resembling an annelid trochophore) and the **Iwata** larva (Fig. 8D) escape from their egg-case and are for a while pelagic. The pilidium and direct types of larvae feed during their larval life, but the Desor and Iwata larvae do not.

Regeneration

Most nemerteans are probably capable of at least some degree of regeneration, a valuable characteristic for animals which so often possess slender and fragile bodies susceptible to mechanical damage. Even violent contractions of the body musculature, such as occur when the animals are handled, can lead to spontaneous fragmentation, especially of the posterior parts (Coe, 1943). The extent to which regeneration can take place depends both upon the species and the part of the body involved. The regeneration of lost or damaged posterior regions seems to be a general feature of nemerteans, but few species are capable of replacing lost anterior parts. Successful regeneration, of whatever form, depends upon at least some part of the central nervous system being present in the body fragment (Coe, 1932).

A few species of lineid and valenciniid Heteronemertea have evolved their regenerative capabilities to such an extent that an asexual multiplicative phase, achieved by spontaneous fragmentation of the body with the subsequent regeneration of each fragment, occurs in their natural life cycle. In these species asexual fission typically takes place during the summer months and alternates with a winter period of normal sexual reproduction (Coe, 1930, 1931). Among the British nemerteans the only species known to possess this ability is *Lineus sanguineus* (Gontcharoff, 1951).

Feeding

Most nemertean species are carnivorous or scavenging and many possess voracious appetites. Although nearly all actively search for their food (Coe, 1943), some species require the close proximity of or even mechanical contact with the prey before commencing feeding behaviour whereas others, such as *Cephalothrix* and *Prostoma*, appear capable of detecting food at a distance, presumably by some chemotactic mechanism (Reisinger, 1926; Jennings & Gibson, 1969). British species whose nutrition and digestion have been studied are *Cephalothrix linearis*, *Cephalothrix rufifrons*, *Lineus ruber*, *Lineus sanguineus*, *Amphiporus lactifloreus* and *Tetrastemma melanocephalum* (Gibson, 1968b; Jennings & Gibson, 1969). Depending upon the condition of the food, nemerteans feed in one of two ways. Active living prey (turbellarians, nematodes, annelids, crustaceans, molluscs or fishes) are caught by the proboscis and either killed or at least partially immobilised by its secretions before being ingested. Inactive or decaying food material, however, is usually swallowed directly without the prior use of the proboscis. When an appropriate organism comes into range, the nemertean forcibly and rapidly ejects its proboscis and coils it around the body of the potential meal. The proboscis is usually aimed fairly accurately, but if it misses it can be quickly withdrawn and re-everted until either the food is caught or moves out of range. In the Palaeo- and Heteronemertea the primary function of the proboscis is to secure the catch, the grip being aided by the viscous secretions and epithelial barbs of the organ, but in the Hoplonemertea the proboscis is initially everted sufficiently to allow the stylet armature to inflict a wound into which toxic secretions can be poured. Once this is achieved the organ is usually everted further and wrapped around the food to hold it.

Many nemerteans ingest their food intact. Anoplan species are capable of dilating their mouth enormously and can swallow organisms almost as large as themselves. Hoplonemerteans, however, are generally incapable of such oral distension and either take small-sized organisms intact or, as in *Amphiporus lactifloreus*, *Nipponnemertes pulcher* and *Tetrastemma melanocephalum* (Jennings & Gibson, 1969; Berg, 1972a; Bartsch, 1973), partially evert their stomach into the prey body and suck out the soft and fluid tissues. Digestion in the carnivorous and scavenging species is effected by a combination of extra- and intracellularly acting enzymes, with proteases forming the major enzymic complement.

The bdellonemertean *Malacobdella grossa* possesses a feeding mechanism that is unique amongst nemerteans (Gibson & Jennings, 1969). *Malacobdella* is an unselective omnivore, predominantly feeding upon small algae, bacteria, protozoans, diatoms and dinoflagellates which it filters from the sea water within the host mantle chamber. Filtering is achieved by a pumping action of the foregut; water and suspended material are first sucked into the barrel-shaped foregut whilst its lining papillae are not intermeshed. The papillae then interlock to form a ciliary net and water is pumped back out. Suitably sized material, including potential food organisms, is then trapped by the cilia and passed back to the rear of the foregut for subsequent passage into the intestine. Unlike the vast majority of ciliary feeding mechanisms found throughout the animal kingdom, which depend upon mucus to trap the food, *Malacobdella* has no mucus involvement at all. The omnivorous diet of this nemertean group is reflected physiologically by its different enzymic complement compared with that of the carnivorous and scavenging species.

Movement

Apart from the pelagic Polystilifera, which either float passively or swim in a slow and sluggish manner, most adult nemerteans move quite freely over hard surfaces, burrow into muddy, sandy or gravelly sediments, or squeeze themselves into rocky crevices or beneath partially embedded rocks or boulders. Benthic forms with dorsoventrally flattened bodies, such as *Cerebratulus*, may, if disturbed, swim actively for a short time by performing vigorous undulatory movements.

Locomotion in adult nemerteans involves the use of both the body wall musculature and the epidermal cilia. Small-sized species (*Oerstedia*, *Tetrastemma*) generally move by a sort of ciliary gliding and a similar type of locomotion is performed by the terrestrial hoplonemertean, *Argonemertes dendyi*. This species secretes copious amounts of mucus, particularly from the cephalic glands, and it effectively 'swims' in this by means of its epidermal cilia (Pantin, 1950). The direction of movement, however, is governed by muscular activity. Other nemerteans creep or glide over objects by a series of posteriorly directed peristaltic muscular waves rippling along the ventral body surface (Coe, 1943); ciliary narcotics do not inhibit this type of locomotion.

All other types of movement involve waves of muscle contractions passing anteriorly or posteriorly along the body and during locomotion the shape and diameter of the animal varies considerably. The bdellonemertean *Malacobdella* often moves in a looping fashion similar to that seen in leeches (Eggers, 1935), alternately using its posterior sucker and oral aperture to attach to the host mantle wall whilst the body is extended or contracted accordingly.

In a few instances the proboscis may be used for rapid locomotion. Species of *Cerebratulus* have been observed burrowing into muddy sediments by means of their proboscis (Wilson, 1900), and several terrestrial nemerteans, including *Argonemertes dendyi*, use the organ to quickly escape from an adverse stimulus (Pantin, 1950). The proboscis is explosively everted to its full length and attaches to the surface by its tip. This is immediately followed by a wave of longitudinal muscle contraction and circular muscle relaxation moving over the body and proboscis, the nemertean 'leaping' forwards as it retracts itself over its own proboscis. That such a rapid type of movement is energy consuming is evidenced by the nemerteans being unable to repeat this behaviour more than once or twice without a rest.

Habitat preferences

Remarkably little is known about the ecology of nemerteans, despite there being many records of particular species from a wide range of habitat types. Kirsteuer (1963a) demonstrated that a loose relationship exists between the nature of the habitat occupied and the systematic position of the species. With the exception of the entozoic Bdellonemertea, each order tends to be associated rather more characteristically with one type of substrate or habitat than another. In general, palaeonemerteans are typically found in finer muddy sediments, hoplonemerteans comprise the dominant nemertean group in algal communities and heteronemerteans, although occurring in a wide range of situations, are often most abundant in coarse muddy sands. Details of the known habitats of each of the British species are given in the Systematic Part later.

Personal observations over a number of years suggest that some species have fairly precise habitat requirements. *Amphiporus lactifloreus* and *Lineus ruber*, for example, are commonly found upon the same shore and not infrequently beneath the same rock or boulder. *Amphiporus*, however, is generally more abundant on moderately clean sandy sediments whilst *Lineus* occurs in greater numbers when the substrate is a muddy sand. Evidence also indicates that the intertidal distribution of these species relative to tidal levels varies on a seasonal basis. Other species apparently occupy a much wider range of habitats.

There are very few observations on the density of nemerteans and, indeed, rocky shore species often exhibit a quite sporadic distribution. Mud- or sand-dwelling forms, however, presumably because they are inhabiting a much more homogeneous type of habitat, have been recorded at densities of from $50/m^2$ (*Prostomatella* sp.: Boyden *et al.*, 1977) up to $500/m^2$ (*Tetrastemma melanocephalum*: R. S. K. Barnes, personal communication).

From what little data are available it is evident that the details of as many habitat features as possible should be recorded during the collection of nemerteans and that there is abundant scope for future investigations into their ecology.

Collection, preservation and identification

The collection of nemerteans requires care if intact specimens are to be obtained and, in general, damaged material is not worth keeping. Several of the larger intertidal and shallow sublittoral species can be found by turning over stones and boulders, sorting through dredge samples or sifting muddy or sandy sediments but, as Kirsteuer (1967a) points out, material so obtained is likely to represent only a minor proportion of the total nemertean fauna inhabiting any particular locality. The following collection procedures, based partly on Kirsteuer's (1967a) suggestions and partly on personal studies, are recommended:

Muddy sediments: Mud samples, whether taken by hand, grab or dredge, are placed in large trays or shallow tanks as soon as possible after collection and gently agitated with clean sea water. Clumped sediments are carefully broken up and any obvious large objects (stones, man-made debris or large invertebrates) removed. The suspended sediment is allowed to settle in an undisturbed state for about 24 hours; Kirsteuer recommends a maximum substrate depth after settlement of 15 cm, with a surface layer of water no more than 2 cm deep. After 24 hours the mud surface and the water/air interface (particularly around the edge of the tray) are carefully examined and any nemerteans removed with a bulb pipette. Examination of the sample is repeated in a further 24–36 hours; if no more nemerteans appear subsamples of the mud, taken from the surface 3–10 mm layer, are dispersed in glass dishes and examined under a dissecting microscope. The main sample is probably exhausted if no specimens appear in the subsamples; if, however, specimens are present the full sample may be re-agitated, allowed to re-settle and examination procedures carried out for a further 24 hours. A second treatment of the sample is recommended only if there are no indications of organic decay commencing.

Sandy sediments: Kirsteuer (1967a) suggests three methods for obtaining nemerteans from sandy sediments.

1. A sample of sand is vigorously stirred with about three times its own volume of sea water. When most of the sand has settled, the water is then poured through a plankton net and any animals retained within the net removed for subsequent examination. This procedure is repeated four or five times for each sample.

2. The sample is treated as above but with the sea water replaced by a 6% $MgCl_2$ solution.

3. 2–3 litre capacity glass cylinders are filled with sand and covered to a depth of about 1 cm with water. After three or four days, if left undisturbed, any nemerteans present will be found in the top 1 cm layer of the sand; subsamples of this layer are removed, examined under a dissecting microscope and any nemerteans removed.

A further method of obtaining interstitial nemerteans (particularly used for species of *Ototyphlonemertes*) was originally described by Corrêa (1958). This involves baiting the surface of wet sand with a morsel of fresh fish meat for about 10–20 minutes. The bait is then removed and the sand below it scooped up into a shallow convex bowl. A small volume of sea water is added, swirled around gently and the sand then allowed to settle to one side by tilting the bowl. Water is then carefully drained away from the settled sand towards the empty side of the container; smaller interstitial nemerteans emerge from the wet sand and can be pipetted out as they move downhill.

Algae: Nemerteans can be induced to emerge from algal fronds and holdfasts by using an oxygen depletion method. A glass jar is about two-thirds filled with algae, and sea water is added to give a minimum depth above the algae of about 5 cm. Kirsteuer recommends keeping any floating plants submerged with weighted wooden sticks. The container, which must not be aerated, is kept dimly lit or in the dark, but one corner is left exposed to natural or artificial light. The respiratory activity of any animals present gradually reduces the level of dissolved oxygen in the water and the dim illumination prevents this from being replenished through algal photosynthesis. At the same time the carbon dioxide concentration in the water increases. The behaviour of any nemerteans present is apparently affected by the increasingly adverse conditions of the water and their usual positively geotactic and negatively phototactic responses are reversed (Kirsteuer, 1967a); as a result they move out of the algae and congregate near the water surface in the lit corner of the jar. This is checked regularly (15–30 minute intervals) and any nemerteans or other invertebrates removed. The sample is mostly exhausted after one to four days. Care must be taken to ensure that the water temperature does not increase too rapidly; if it does the rate of oxygen depletion may be so rapid that animals become asphyxiated before they can reach the water surface.

Rocks, coral fragments, samples of colonial sedentary animals: The oxygen depletion method described above is equally effective with other types of substrate which might harbour nemerteans. The surface of the sample is examined for larger specimens and fragments of more than about 10 cm diameter are broken up before immersion in clean sea water. Kirsteuer recommends that the most convenient type of container is a rectangular glass tank of about 30 litres capacity which is taller than it is wide.

The identification of nemerteans ideally depends upon the careful examination of living material, followed by histological studies of paraffin wax sections. The following procedures are recommended:

Examination of living material: This is most easily done with anaesthetised individuals. Kirsteuer (1967a) regards urethane and chloral hydrate as very satisfactory narcotics, but the author uses a 7.5–8.0% solution of $MgCl_2$ in distilled water (2.0–2.5% for freshwater species). Nemerteans immersed in this fluid are usually anaesthetised and relaxed in 5–20 minutes depending upon the species. $MgCl_2$, which acts as a muscle relaxant, has the advantage that after examination animals can be returned to fresh sea water and will fully recover within about one hour. Details which should be recorded are:

1. General body shape (e.g. length, maximum width, whether tapering posteriorly or of uniform width throughout, whether rounded or dorso-ventrally flattened in the intestinal regions).
2. Shape of cephalic region (e.g. pointed, rounded, with a distinct cephalic lobe).
3. Caudal cirrus (present or absent).
4. Cephalic slits or grooves (e.g. lateral longitudinal, dorsal oblique).
5. Colour (background colour, differences between anterior and posterior regions or dorsal and ventral surfaces, arrangement of any distinct pattern).
6. Eye number and distribution or absence of eyes (careful observation is needed if the cephalic region is darkly pigmented as the colour may hide the eyes).

Many species are almost transparent or only a very pale colour; in these, internal organs such as the rhynchocoel, proboscis, intestine, gonads and cerebral ganglia can often be distinguished through the dorsal body surface. Appropriate details on these structures, together with other observations made on the living nemerteans, are best recorded on a fully annotated drawing.

It is often advantageous to lightly flatten the anaesthetised individual before examining it under a dissecting microscope; in this way internal structures are more clearly revealed (this procedure is especially useful for demonstrating the eyes and the proboscis armature of hoplonemerteans). The required degree of flattening can be achieved by placing the nemertean between two glass slides and applying pressure; care must be taken to avoid excessive pressure which may burst the animal.

Histological studies: Although several nemertean species can be reliably identified just on the basis of their external features, this is only possible if the animals are seen alive. Often specimens, such as those obtained from dredge samples, are already preserved and in such cases histological studies are essential. Indeed, these procedures are advisable in every instance if the collector possesses the necessary time, enthusiasm and facilities, whether commencing with living or preserved specimens. The routines outlined below are those regularly employed by the author; other individuals may prefer to adopt their own modifications.

1. *Fixation:* Bouin's or Susa's fixatives are, in general, ideal for nemerteans, although Bouin's fluid is not suitable for the stylet apparatus of hoplonemerteans. Formalin or alcohol, widely used by museums, etc., for the preservation of bulk samples, are not recommended for nemerteans although samples so preserved can still be examined histologically. It is best to apply the fixative to anaesthetised specimens, keeping them as straight as possible during fixation.

2. *Sectioning:* The internal morphology is more easily followed and understood from transverse than from longitudinal sections. Small specimens (10 mm or less in length) may be completely sectioned (56 °C melting point Paraffin wax is recommended), but with larger individuals the size makes complete sectioning impracticable. In these cases the whole of the anterior end (up to the rear of the foregut) should be sectioned, together with representative 'segments' taken from the remainder of the body at more or less regular intervals (e.g. every 5 cm in an animal with a total length of 30–40 cm).

3. *Staining:* Although sometimes inconsistent in application, Mallory's trichrome has been found to be a useful general stain for distinguishing between the various body tissues and organ systems of nemerteans. Other staining procedures, such as the Azan or Masson methods, provide comparable results but take much longer to use. It is often beneficial to stain a few sections with 1% Alcian blue, counterstained with eosin or acid fuchsin; mucus-secreting glands of the epidermis, dermis, cephalic region and gut are shown up particularly well by this procedure.

Classification of the Nemertea

The classification of the Nemertea (= Nemertinea, Nemertini or Rhynchocoela) is based upon internal morphology. There is a general agreement on the higher classification, the phylum being divided into two classes, the Anopla and Enopla, each of which is further subdivided into two orders, as given in Gibson (1972). The revision proposed by Iwata (1960), involving the introduction of a new anoplan order, the Archinemertea, to accommodate the cephalothricid palaeonemerteans, and relegating the enoplan order Bdellonemertea to a suborder of the Hoplonemertea, has not subsequently been adopted. Familial and generic groupings, conversely, are far less securely established and, as indicated in the Introduction, urgently require revision.

The higher classification

Phylum NEMERTEA

Unsegmented, bilaterally symmetrical, acoelomate invertebrates with a gut possessing separate mouth and anus, a blood vascular system, and an eversible muscular proboscis situated dorsal to the gut in an enclosed tubular chamber, the rhynchocoel.

Class ANOPLA Mouth below or posterior to the cerebral ganglia; central nervous system situated within the body wall (epidermis, dermis or body wall musculature); proboscis not differentiated into three regions and either not armed or provided with large numbers of rhabdite-like epithelial barbs.

Order PALAEONEMERTEA
Layers of body wall musculature either two (outer circular, inner longitudinal) or three (outer circular, middle longitudinal, inner circular); dermis of hyaline connective tissue; central nervous system either in inner longitudinal musculature or external to body wall muscles.

Order HETERONEMERTEA
Body wall musculature primarily three layered (outer longitudinal, middle circular, inner longitudinal), sometimes with additional thin inner circular and/or outer oblique layers; dermis well developed and normally composed of fibrous connective tissues and gland cells; central nervous system situated between outer longitudinal and middle circular muscle layers.

Class ENOPLA Mouth anterior to cerebral ganglia; central nervous system internal to body wall musculature, which is two layered (outer circular, inner longitudinal); proboscis usually regionally differentiated and armed with one or more needle-like stylets.

Order HOPLONEMERTEA

Proboscis armed with one or more stylets; intestine straight, mostly with paired lateral diverticula; no posterior ventral sucker.

Suborder MONOSTILIFERA

Proboscis armature a single central stylet carried on a large cylindrical basis.

Suborder POLYSTILIFERA

Proboscis armature a pad or shield bearing numerous small stylets.

Tribe REPTANTIA Crawling or burrowing forms; rhynchocoel with caecal outgrowths; cerebral organs and nephridial system present.

Tribe PELAGICA Bathypelagic forms found swimming or floating in deep water; rhynchocoel without caeca; cerebral organs and nephridial system absent.

Order BDELLONEMERTEA

Proboscis not armed with stylets; intestine sinuous and without lateral diverticula; foregut barrel-shaped and lined with tracts of papillae; with a posterior ventral sucker.

The classification of the British nemerteans

Phylum NEMERTEA
Class ANOPLA
Order PALAEONEMERTEA
Family Carinomidae
Carinoma armandi (McIntosh, 1875)
Family Cephalothricidae
Cephalothrix linearis (Rathke, 1799)
Cephalothrix rufifrons (Johnston, 1837)
Procephalothrix filiformis (Johnston, 1828–29)
Family Tubulanidae
Carinesta anglica Wijnhoff, 1912
Tubulanus albocapitatus Wijnhoff, 1912
Tubulanus annulatus (Montagu, 1804)
Tubulanus banyulensis (Joubin, 1890)
Tubulanus inexpectatus (Hubrecht, 1880)
Tubulanus linearis (McIntosh, 1873–74)
Tubulanus miniatus (Bürger, 1892)
Tubulanus nothus (Bürger, 1892)
Tubulanus polymorphus Renier, 1804
Tubulanus superbus (Kölliker, 1845)
Order HETERONEMERTEA
Family Baseodiscidae
Baseodiscus delineatus (Delle Chiaje, 1825)
Oxypolia beaumontiana Punnett, 1901
Family Lineidae
Cerebratulus alleni Wijnhoff, 1912
Cerebratulus fuscus (McIntosh, 1873–74)
Cerebratulus marginatus Renier, 1804
Cerebratulus pantherinus Hubrecht, 1879
Cerebratulus roseus (Delle Chiaje, 1841)
Euborlasia elizabethae (McIntosh, 1873–74)
Lineus acutifrons Southern, 1913
Lineus bilineatus (Renier, 1804)
Lineus lacteus (Rathke, 1843)
Lineus longissimus (Gunnerus, 1770)
Lineus ruber (Müller, 1774)
Lineus sanguineus (Rathke, 1799)
Lineus viridis (Müller, 1774)
Micrella rufa Punnett, 1901
Micrura aurantiaca (Grube, 1855)
Micrura fasciolata Ehrenberg, 1831
Micrura lactea (Hubrecht, 1879)
Micrura purpurea (Dalyell, 1853)

Micrura rockalliensis Dollfus, 1924
Micrura scotica Stephenson, 1911
Family Poliopsiidae
Poliopsis lacazei Joubin, 1890
Family Valenciniidae
Valencinia longirostris Quatrefages, 1846
Class ENOPLA
Order HOPLONEMERTEA
Suborder Monostilifera
Family Amphiporidae
Amphiporus allucens Bürger, 1895
Amphiporus bioculatus McIntosh, 1873–74
Amphiporus dissimulans Riches, 1893
Amphiporus elongatus Stephenson, 1911
Amphiporus hastatus McIntosh, 1873–74
Amphiporus lactifloreus (Johnston, 1827–28)
Family Carcinonemertidae
Carcinonemertes carcinophila (Kölliker, 1845)
Family Cratenemertidae
Nipponnemertes pulcher (Johnston, 1837)
Family Emplectonematidae
Emplectonema echinoderma (Marion, 1873)
Emplectonema gracile (Johnston, 1837)
Emplectonema neesii (Örsted, 1843)
Nemertopsis flavida (McIntosh, 1873–74)
Family Prosorhochmidae
Argonemertes dendyi (Dakin, 1915)
Oerstedia dorsalis (Abildgaard, 1806)
Oerstedia immutabilis (Riches, 1893)
Oerstedia nigra (Riches, 1893)
Prosorhochmus claparedii Keferstein, 1862
Family Tetrastemmatidae
Prostoma graecense (Böhmig, 1892)
Prostoma jenningsi Gibson & Young, 1971
Tetrastemma ambiguum Riches, 1893
Tetrastemma beaumonti (Southern, 1913)
Tetrastemma candidum (Müller, 1774)
Tetrastemma cephalophorum Bürger, 1895
Tetrastemma coronatum (Quatrefages, 1846)
Tetrastemma flavidum Ehrenberg, 1831
Tetrastemma helvolum Bürger, 1895
Tetrastemma herouardi (Oxner, 1908)
Tetrastemma longissimum Bürger, 1895
Tetrastemma melanocephalum (Johnston, 1837)
Tetrastemma peltatum Bürger, 1895

Tetrastemma quatrefagesi Bürger, 1904
Tetrastemma robertianae McIntosh, 1873–74
Tetrastemma vermiculus (Quatrefages, 1846)
Suborder Polystilifera
Tribe Reptantia
Family Drepanophoridae
Punnettia splendida (Keferstein, 1862)
Family Paradrepanophoridae
Paradrepanophorus crassus (Quatrefages, 1846)
Order BDELLONEMERTEA
Family Malacobdellidae
Malacobdella grossa (Müller, 1776)

Key to the species of British nemerteans

Note: The key is based primarily upon features distinguishable in living material and its use requires neither a detailed knowledge of the group nor any equipment more elaborate than a reasonably powerful hand lens or a low-power binocular microscope. Several species, however, cannot be reliably identified by their external characters alone; for these forms internal distinguishing features, requiring histological studies for their determination, are provided in italics within parentheses. Inadequately described species are indicated with an asterisk.

1. Freshwater or terrestrial nemerteans **2**

Marine or estuarine nemerteans **4**

2(1). Terrestrial, found in damp and shaded habitats under logs, fallen branches, beneath stones or among decaying leaves and moss; cream or pale yellowish with a longitudinal dark brown stripe running along each dorsolateral margin and extending from behind the four groups of eyes to the tail; ventral surface paler, without stripes .... *Argonemertes dendyi* (p. 140)

Freshwater, in ponds or rivers on aquatic plants or on the surface of mud; uniformly reddish-brown, orange or brown as adults, occasionally greenish, juveniles white to pale yellow or straw coloured; eyes 4–6 depending upon age and size (Genus: *Prostoma*) **3**

3(2). (*With distinct, ciliated oesophagus; cephalic glands reaching cerebral ganglia; proboscis with 9–10 nerves; rhynchodaeum with well developed longitudinal muscles*) .... *Prostoma graecense* (p. 152)

(*With distinct but non-ciliated oesophagus; cephalic glands not posteriorly reaching cerebral ganglia; proboscis with 11 nerves; rhynchodaeum with weakly developed longitudinal muscles*) .. .. *Prostoma jenningsi* (p. 154)

4(1). Nemerteans either living commensally in the mantle chamber of bivalve molluscs or parasitically on the egg masses or between the gill filaments of crabs **5**

Free living nemerteans **6**

5(4). Small, slender nemerteans, yellowish, orange, pale reddish, rose pink or bright brick red in colour; without a posterior sucker; with two small eyes; parasitic on crabs .... *Carcinonemertes carcinophila* (p. 122)

Flattened, leech-like nemerteans inhabiting the mantle cavity of bivalve molluscs; with a posterior ventral sucker; without eyes; immature individuals white, cream or pale grey, mature adults with cream (males) or olive-green or yellowish-green (females) gonads .... *Malacobdella grossa* (p. 182)

6(4). With a distinctly marked colour pattern consisting of longitudinal stripes and/or transverse bands, one or more patches of cephalic pigment which contrast markedly with the remaining body colour, or mottled, variegated or speckled patterns, at least on the dorsal surface **7**

Without a distinct colour pattern, body usually more or less uniformly coloured, sometimes paler on ventral surface or in anterior and/or posterior regions **33**

7(6). Colour pattern consisting of well defined white longitudinal stripes and transverse bands on a reddish background **8**

Colour pattern of a different form **11**

8(7). Head white or colourless, with or without a pair of dorsal black crescentic pigment patches near the anterior cephalic margins **9**

Head red, orange or reddish-brown, with a mid-dorsal longitudinal white stripe which extends the full length of the body **10**

9(8). Head without black pigment patches, body reddish-brown dorsally, paler ventrally, with a distinct median dorsal stripe and transverse bands, lateral stripes incomplete and indistinct, no mid-ventral stripe .... **Tubulanus albocapitatus* (p. 58)

Head with black pigment patches, body dorsally dark reddish-brown, ventrally pale orange-brown to yellowish, with distinct dorsal median and lateral stripes and transverse bands, mid-ventral stripe usually present but inconspicuous .... **Tubulanus nothus* (p. 66)

10(8). Body a vivid brick-red, orange or brownish-red, with distinct median dorsal and lateral stripes and transverse bands, without mid-ventral stripe .... *Tubulanus annulatus* (p. 60)

Body a dark reddish-brown, scarlet or cherry-red with distinct median dorsal, median ventral and lateral stripes and transverse bands .... *Tubulanus superbus* (p. 69)

11(7). Colour pattern consisting of five or more dorsal longitudinal stripes which may be interrupted or irregularly fused with each other, without transverse bands **12**

Colour pattern consisting of three or less dorsal longitudinal stripes or of a different form **13**

12(11). Dorsal surface coloured buff or reddish-brown, with five distinct longitudinal stripes of white, grey or pale pink; lateral margins whitish; ventral surface yellowish-pink or pinkish-brown without distinct markings; eyes arranged on head in four longitudinal rows ....*Punnettia splendida* (p. 178)

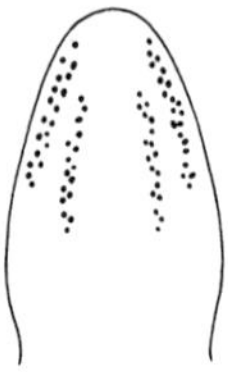

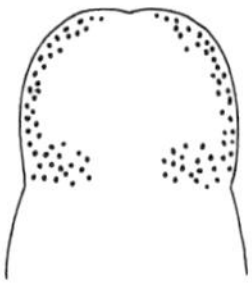

Dorsal surface a dull yellowish-fawn to light brown, marked with 5–12 reddish-brown interrupted longitudinal stripes of variable width and outline, adjacent stripes often fusing with each other; ventral surface generally with fewer and paler stripes; eyes not arranged in four longitudinal rows, mostly distributed along margins of head but extending medially near back of head to form two large dorsolateral groups ....*Baseodiscus delineatus* (p. 70)

13(11). Colour pattern consisting of one to three dorsal longitudinal stripes, either without transverse bands or with a single band at the rear of the head **14**

Colour pattern of a different form **18**

14(13). General body colour pale, translucent, yellowish, orange or pinkish-brown, with four distinct eyes **15**

General body colour dark brown, chocolate brown or reddish-brown, either with four eyes, which are sometimes masked by the cephalic pigmentation, or without eyes **17**

15(14). With single mid-dorsal stripe **16**

With band of brown pigment encircling the body at the rear of the head, sometimes ventrally incomplete, and a pair of dorsolateral brown longitudinal stripes extending from the rear of the transverse band towards the tail; with a median dorsal white stripe and with variably developed white pigment patches on the head between the two pairs of eyes ....*Tetrastemma robertianae* (p. 174)

16(15). Body more or less transparent, light flesh coloured, with a single median dorsal stripe of dark wine red extending the full body length .... *Tetrastemma herouardi* (p. 166)

Body yellowish, not transparent, dorsally marked with chocolate brown or orange-red pigment specks which are strongly concentrated into a median dorsal line extending the full body length .... **Oerstedia immutabilis* (p. 147)

17(14). With four eyes; body brown or reddish-brown with a single median dorsal stripe of pale yellow, cream or dirty white which may extend the full body length continuously or be irregularly interrupted .... *Oerstedia dorsalis* (p. 144)

Without eyes; body colour variable but typically a rich reddish- or chocolate brown, sometimes paler ventrally, marked with two slender white or pale yellow longitudinal stripes dorsally which flank a narrow median stripe of reddish-brown; the white stripes are usually more widely separated on the .... *Lineus bilineatus* (p. 86)

18(13). Colour pattern restricted to one or more patches of pigment on the head which contrast markedly with the remaining body colour **19**

Colour pattern of a different form **25**

19(18). With four distinct eyes arranged at the corners of a square or rectangle (*note*: the eyes may be partially obscured by the cephalic pigmentation); without a caudal cirrus **20**

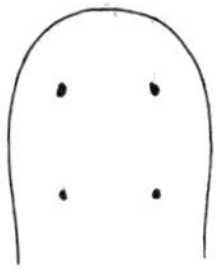

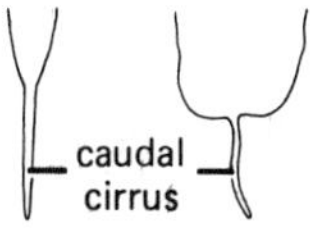

Without eyes; with a caudal cirrus **24**

caudal cirrus

20(19). Head with a single patch of pigment dorsally of variable shape **21**

Head with two longitudinal streaks of dark brown extending between the anterior and posterior eyes on each side; general body colour dull whitish, salmon, pink, pale orange or apricot yellow, intestinal regions sometimes pale green ....
.... **Tetrastemma vermiculus* (p. 176)

21(20). Cephalic pigment patch quadrangular in shape or nearly so, usually covering at least one pair of eyes **22**

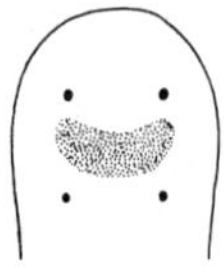

Cephalic pigment patch distinctly crescentic in shape, located midway between and not covering the eyes **23**

22(21). General body colour yellowish or yellow-green, cephalic pigment patch black or dark brown, covering most of the head area between the eyes .... *Tetrastemma melanocephalum* (p. 168)

General body colour brownish-yellow, head almost colourless apart from a bright to dark red or reddish-brown pigment patch which rarely covers more than half the area enclosed by the eyes .... **Tetrastemma longissimum* (p. 167)

23(21). General body colour fawn to deep brown, intestinal region sometimes greenish; with brown or black cephalic pigment patch developed into three anterior points ....
.... **Tetrastemma peltatum* (p. 170)

General body colour pale yellowish-green, light green or light brownish-green; cephalic pigment patch of dark brown or black without a median anterior point ....
.... **Tetrastemma coronatum* (p. 162)

24(19). Body coloured a bright brick red dorsally, white or pinkish ventrally; on the head a white patch separates an anterior dorsal cephalic spot of reddish, violet or brownish pigment which is very variable in both size and shape, exceptionally it is completely missing; extreme tip of head white ....
.... *Micrura aurantiaca* (p. 100)

Body a dark, rich, iridescent purplish-brown dorsally, similar in colour but generally paler ventrally; head marked by a dorsal transverse band of brilliant yellow or yellowish-white which is sometimes divided into two or, rarely, is absent; in front of the band the tip of the head is whitish or translucent, behind the band there is usually a median accumulation of white granules .... *Micrura purpurea* (p. 105)

25(18). Colour pattern consisting of more or less regularly arranged transverse bands, without distinctly marked longitudinal stripes **26**

With a distinctly mottled, variegated or speckled colour pattern, at least on the dorsal surface **28**

26(25). With two or more eyes which are usually at least partly obscured by the general body colour **27**

Without eyes, head white or pinkish with two small dorsal patches of black pigment near its anterior margins; body dorsally bright cherry-red or reddish-brown to muddy green, ventrally pinkish or yellowish, with up to about 18 white transverse bands encircling the body ....
.... **Tubulanus banyulensis* (p. 61)

27(26).

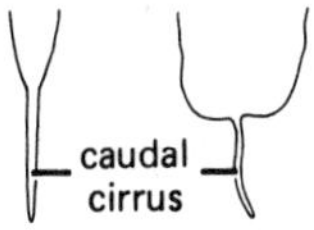

caudal cirrus

With 3–12 small eyes (fewer in juvenile worms) arranged in a row near each cephalic margin; with a caudal cirrus; body usually coloured a rich reddish-brown, sometimes yellowish-brown, greenish-brown or reddish-violet, marked with white transverse bands dorsally, ventrally paler and without obvious transverse bands; the head normally has a dorsal patch of white pigment of variable shape, but this may be absent ....
.... *Micrura fasciolata* (p. 102)

With four eyes arranged approximately in a square; without a caudal cirrus; general body colour varying from pale yellowish-brown to cinnabar-brown or reddish-brown, paler ventrally, marked dorsally with transverse bands of dark brown, chestnut-brown or brownish-yellow ....
.... *Oerstedia dorsalis* (p. 144)

28(25). With four or more eyes **29**

Without eyes **31**

29(28). With four eyes arranged approximately in a square, although the eyes may be partially obscured by the body colour **30**

With 20–30 or more small eyes on each side of the head arranged into antero- and posterolateral groups; general body colour pale yellowish-brown, straw or flesh, marked dorsally with irregular dark brown pigment specks or streaks which give a speckled or longitudinally mottled appearance, ventral surface pale pinkish-white or flesh coloured .... *Emplectonema neesii* (p. 134)

30(29). Eyes reddish; body dorsally dark brown to almost black due to the closely reticulated arrangement of the pigment specks, sometimes with a vaguely defined median dorsal yellowish stripe, ventral surface pale yellowish .... **Oerstedia nigra* (p. 148)

Eyes black; body pale yellowish-brown to reddish-brown or reddish-orange, dorsally speckled with brilliant white granules and marked with irregular patches of dark brown to brownish-yellow pigment, sometimes with a thin line of brown pigment along each side of the body, ventral surface paler than dorsal .. .. *Oerstedia dorsalis* (p. 144)

31(28). Body rounded, not distinctly flattened and without sharp lateral margins; without a caudal cirrus **32**

Body distinctly flattened, with sharp lateral margins; with a caudal cirrus; dorsally marked by a mottled pattern of irregular dirty-green, brown, yellow and white pigment patches which are especially evident in the anterior regions .... **Cerebratulus pantherinus* (p. 80)

32(31).

lateral cephalic slit

With lateral horizontal cephalic slits on the head, the head obviously narrower than the succeeding body regions; body bulky and with a distinctly wrinkled surface; head mostly white or pale yellowish, speckled with olive-green or brown pigment flecks, remaining body dark brown or reddish-brown irregularly marbled with white, yellow, light brown and olive-green pigment patches, more or less regularly marked with indistinctly defined transverse bands of pale pink or dirty white .... **Euborlasia elizabethae* (p. 82)

Without lateral horizontal cephalic slits on the head, the head not narrower than the adjacent body regions; body neither bulky nor with a distinctly wrinkled surface; dorsally coloured a brilliant vermilion or reddish-brown, speckled with white, ventral surface somewhat paler, tip of head colourless .... ***Tubulanus inexpectatus* (p. 62)

33(6). 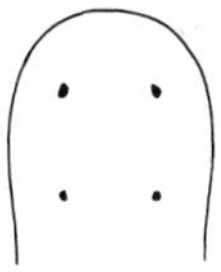 With four distinct eyes arranged at the corners of a square or rectangle; without lateral horizontal cephalic slits; small (usually less than 40 mm long), slender nemerteans, bodies translucent with the internal organs visible or body coloured in pale shades of pink, yellow, green or orange **34**

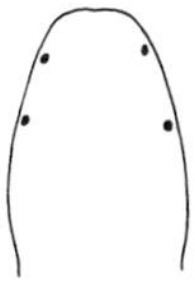

With two eyes, more than four eyes or without eyes (*note*: young individuals of some species may have only four eyes even though more are present in adults; in these instances the four eyes are not arranged at the corners of a square or rectangle but are situated at the cephalic margins) with or without horizontal lateral cephalic slits **42**

34(33).  Head distinctly diamond-shaped and wider than adjacent body regions; eyes either large, distinct and of approximately equal size or irregularly shaped and with anterior pair significantly bigger than the posterior pair **35**

Head either not wider than adjacent body regions or, if slightly wider, not obviously diamond-shaped **36**

35(34). Eyes typically coloured brown, sometimes black, and irregularly shaped, the anterior pair at least twice the size of the posterior pair; body an overall pale yellow but usually marked dorsally with a variable amount of reddish-brown pigmentation .... ***Tetrastemma ambiguum* (p. 156)

Eyes distinct, large and more or less of an equal size; dorsally coloured a uniform reddish-brown but head, ventral surface and lateral margins pale yellowish .... ***Tetrastemma cephalophorum* (p. 161)

36(34). Body stout, cylindrical, distinctly wider towards the bluntly rounded tail; colour creamish-white, occasionally with a faint pink tinge; under a microscope irregularly distributed and shining epidermal glands are usually evident .... **Tetrastemma beaumonti* (p. 158)

Body rather flattened, with a more or less uniform width throughout or gradually tapering posteriorly **37**

37(36). Anterior pair of eyes distinctly larger than the posterior pair .. **38**

Anterior and posterior pairs of eyes of approximately the same size **39**

38(37). Body slender, filiform, with both ends tapered; distance between anterior and posterior pairs of eyes greater than distance between eyes of the same pair; colour a uniform white, yellowish, pink, pale peach or reddish-brown, with pale lateral margins and translucent snout; blood system may appear reddish; (*rhynchocoel at most about half body length*; *nervous system without neurochords*; *intestinal caecum short, without anterior diverticula*) .... *Nemertopsis flavida* (p. 138)

Body with a more or less uniform width throughout; head sometimes with terminal notch, giving a rather bilobed appearance, and slightly wider than adjacent body regions; colour typically pale yellow to orange; (*rhynchocoel extends the full length of the body*; *nervous system with neurochords*; *intestinal caecum with anterior diverticula*) .... *Prosorhochmus claparedii* (p. 150)

39(37). Tip of head, between the anterior pair of eyes, and anal region with dense accumulations of gland cells which appear as shining whitish masses; body coloured a bright pale honey-yellow, head pale yellow .... **Tetrastemma helvolum* (p. 164)

Not as above **40**

40(39). Colour a transparent yellowish; proboscis with four accessory stylet pouches, visible when animal is lightly flattened and viewed under a microscope .... **Tetrastemma quatrefagesi* (p. 172)

Colour white, pink or brown or, if yellowish, with only two accessory stylet pouches in the proboscis **41**

41(40). Eyes moderately large and distinct, reddish-brown, dark brown or black; body commonly with a greenish hue, but colour extremely variable (shades of yellow, brownish, orange, greyish-yellow, greenish-yellow), especially in the intestinal regions; sometimes with an opaque whitish patch between the eyes and/or with a slender median longitudinal white streak extending along the back; cephalic furrows colourless or brown .... **Tetrastemma candidum* (p. 159)

Eyes very small but usually distinct; body typically coloured bright pink, but smaller individuals may be pale yellow or reddish and appear more or less transparent; lateral margins translucent .... **Tetrastemma flavidum* (p. 163)

42(33). Head with distinct longitudinal lateral cephalic slits **43**

Head without lateral cephalic slits, either without slits or furrows at all or with dorsal or ventral longitudinal, transverse or oblique grooves **56**

43(42). With a caudal cirrus **44**

Without a caudal cirrus **51**

44(43). Body in intestinal region broad and flattened, with distinct thin lateral margins **45**

Body either not distinctly flattened or, if so, without distinct thin lateral margins **48**

45(44). Head conspicuously swollen and pear- or fig-shaped; body a light flesh colour with a white cephalic tip; (*body wall musculature without diagonal layer*; *dermis without distinct connective tissue zone*, *dermal glands distributed amongst body wall outer longitudinal muscle fibres*)............ **Cerebratulus alleni* (p. 75)

Head not as above; body colour differs from above **46**

46(45). With small eyes distributed along the cephalic margins, the eyes sometimes not easily distinguished because of the body colouration **47**

Without eyes; head yellowish with cerebral ganglia shining reddish at its rear; pink or light flesh coloured in the foregut regions, posteriorly yellowish or with a faint reddish hue; in ripe females the orange-brown colour of the eggs may give the body a darker appearance; lateral margins colourless and usually quite conspicuous ....**Cerebratulus roseus* (p. 81)

47(46). General colour pale yellowish, greyish-brown or pinkish, dorsally speckled lightly with red, brown or greenish-grey, often the anterior regions appear more darkly coloured; body commonly marked by pale transverse furrows or wrinkles; with 4–13 eyes on each side of the head .... *Cerebratulus fuscus* (p. 76)

Colour typically a uniform greyish-brown with whitish or transparent lateral margins, but dark greyish-green, slate-blue or dull brown examples are reported; eyes present but indistinct .... *Cerebratulus marginatus* (p. 78)

48(44). Without eyes **49**

With 5–8 eyes on each side of the head, the anterior eye of each row larger and more distinct than the rest; dorsal colour a light brown anteriorly, with purplish tinge, but in the intestinal region only the gut and its diverticula are similarly coloured, the remainder of the body pale; margins of head, mouth and body white, ventral surface whitish; head with small reddish patch near the tip between the anterior pair of eyes .... **Micrura scotica* (p. 107)

49(48). Colour typically an overall milk white, but intestinal contents may impart a pale yellowish or brownish to flesh-pink hue and occasional examples may be more uniformly tinged a pale rose-pink; under a microscope the epidermal surface appears covered with opaque white flakes due to the large size of many gland cells .... *Micrura lactea* (p. 104)

Colour different from above **50**

50(49). Body a uniform bright vermilion red, shading into yellowish near the tip of the head; cerebral ganglia appear bright red; intestine and its diverticula brown .... *Micrella rufa* (p. 98)

Colour in life not known; after preservation in alcohol body a uniform blackish-brown .... **Micrura rockalliensis* (p. 106)

51(43). With eyes distributed along the lateral cephalic margins **52**

Without eyes; head acutely pointed; colour either pure white anteriorly, gradually becoming pink or light red to brownish-purple posteriorly, or coral red in front and pink behind .... **Lineus acutifrons* (p. 84)

52(51). Colour a distinct reddish-brown throughout **53**

Colour different from above... **54**

53(52). With 2–8 eyes on either side of the head; colour typically light to dark reddish-brown, paler ventrally, but sometimes with violet, greenish-red or yellowish-brown tinge; when mechanically irritated the nemerteans contract without coiling into a spiral. .... *Lineus ruber* (p. 92)

With 4–6 eyes on either side of the head; colour varying from light reddish-brown to dull mid-brown, often posteriorly and ventrally somewhat paler; when mechanically irritated the nemerteans characteristically contract into a tight spiral coil .... *Lineus sanguineus* (p. 94)

54(52). With 6–15 eyes on each side of the head; head pale pink or reddish, fading posteriorly to a uniform white or pale creamish-yellow; tip of head and tail translucent .... *Lineus lacteus* (p. 87)

Colour not as above **55**

55(54). With 10–20 small eyes on each side of the head; colour varying from dark olive-brown or rich chocolate brown to blackish-brown or black, usually with a flickering purple iridescence due to the activity of epidermal cilia; body may appear faintly streaked with pale longitudinal lines or creases; when handled, the nemerteans produce copious amounts of a sticky, rather pungent, mucus ....*Lineus longissimus* (p. 88)

With 2–8 eyes on each side of the head; colour pale to dark olive-green or greenish-black; gonopores white in sexually mature individuals.................................... *Lineus viridis* (p. 96)

56(42). With eyes **57**

Without eyes **67**

57(56). With less than six eyes **58**

With more than six eyes, usually more than 20 .... **59**

58(57). With two large eyes near the tip of the head; body dorsally a dull orange or pale brown colour, tending to reddish anteriorly and especially on the head, ventral surface paler; milk-white, creamish-white, rose-red or dark green colour varieties are recorded .... **Amphiporus bioculatus* (p. 114)

With more than two eyes, the anterior pair near the tip of the head some distance in front of the remaining eyes; colour an overall bright yellow with paler, whitish, lateral margins; when contracted the body appears more orange in colour; cerebral ganglia appear distinctly reddish .... **Amphiporus elongatus* (p. 116)

59(57). Head without median dorsal longitudinal furrow or, if one is present, body distinctly flattened and not deeply wrinkled when contracted **60**

Head with a deep median dorsal furrow and bearing about 80 small eyes on the dorsolateral margins; body bulky, surface very wrinkled when contracted; tip of head colourless; body bright pink to greyish-red anteriorly, yellowish posteriorly ....*Poliopsis lacazei* (p. 108)

60(59). Eyes arranged in four distinct longitudinal rows on the head **61**

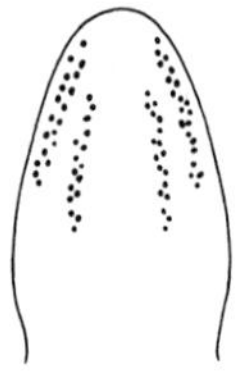

Eyes arranged in other ways **62**

61(60). Body distinctly flattened throughout its length; head tapered and narrower than adjacent body regions; general colour light to dark brown dorsally, shading to dull orange near the pale lateral margins, ventral surface light rose to pinkish-white .... *Paradrepanophorus crassus* (p. 180)

Body not distinctly flattened; head oval and slightly wider than adjacent body regions; head and anterior body pale yellowish, rest of body light salmon-pink; cerebral ganglia distinctly pinkish .... **Amphiporus allucens* (p. 112)

62(60). Eyes arranged in two distinct groups on either side of the head **63**

Eyes arranged in a single continuous row, sometimes dispersed, on either side of the head **64**

63(62). Eyes distinct; body colour very variable, generally a dull pink or dirty white with paler head, tail and lateral margins, cerebral ganglia reddish; mature females often orange or reddish, males grey or brown in the intestinal regions; gut contents may appear green, blackish or dark grey; does not produce copious amounts of mucus when irritated ....*Amphiporus lactifloreus* (p. 120)

Eyes to some extent obscured by dorsal cephalic pigmentation; body dorsally a dark dull olive-green, dark blue-green, greenish-brown or greyish-green colour, sometimes with a blue-grey iridescence, ventral surface pale greyish-yellow, grey-green or dirty yellow-white; when irritated the nemerteans secrete copious amounts of a whitish mucus .... *Emplectonema gracile* (p. 132)

64(62). 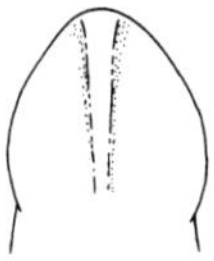 Head a rounded shield shape with a median dorsal ridge or furrow **65**

Head oval or diamond-shaped, without median ridge or furrow **66**

65(64). Body pink, yellowish-brown, bright red, pale brown or dark greyish-brown, the ventral surface and head paler; head with a distinct pale median dorsal ridge ..... **Amphiporus hastatus* (p. 118)

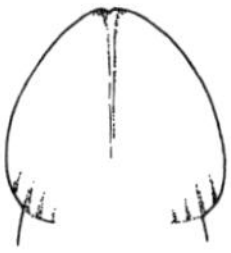

Body dorsally brown, red or pink, lateral margins and ventral surface much lighter; head with a distinct median dorsal swelling flanked by longitudinal furrows; anterior transverse cephalic furrow with secondary slits .... *Nipponnemertes pulcher* (p. 126)

66(64). Head oval; body light to dark salmon-pink, orange, yellowish-brown or light reddish-brown, intestinal region often darker; epidermis without sickle-shaped spicules; (*rhynchocoel more or less the full length of the body*; *cephalic glands not well developed*) .... *Amphiporus dissimulans* (p. 115)

Head distinctly diamond-shaped; body pale salmon, yellowish-red or orange-red, generally paler in the posterior third; young individuals may be white or colourless; epidermis containing large numbers of minute transparent and refractile sickle-shaped spicules; (*rhynchocoel at most half the length of the body*; *cephalic glands well developed*) .... *Emplectonema echinoderma* (p. 130)

67(56). With distinct transverse furrow encircling the rear of the head; body flattened when extended, otherwise anteriorly cylindrical apart from the flat and pointed head; anterior half of body milk white, intestinal region tinged brownish or pale rose-red; (*body wall and proboscis musculature comprising outer longitudinal, middle circular and inner longitudinal layers*) .... *Oxypolia beaumontiana* (p. 72)

Without distinct transverse furrow on the head; (*body wall musculature either with two or three principal layers, the outermost composed of circular fibres or, if arranged as above, proboscis with only two muscle layers*) **68**

68(67). Body slender, with mouth placed far behind cerebral ganglia; head very long and pointed **69**

Mouth located close to cerebral ganglia **71**

69(68). Nemerteans when irritated may contract into a tight knot but do not coil spirally; (*no inner circular muscle layer in foregut region*) (Genus: *Cephalothrix*) **70**

Nemerteans contract in a tight spiral when disturbed; head long, white or translucent, remainder of body yellowish-white to orange, often darker posteriorly; (*with an incomplete inner circular muscle layer in foregut region*) .... *Procephalothrix filiformis* (p. 56)

70(69). Body coloured white, cream or pale greyish and translucent, sometimes tinged yellowish anteriorly; gut contents may variably colour the intestinal region; head without orange or reddish pigmentation ....*Cephalothrix linearis* (p. 52)

Body coloured whitish or translucent, with a variable amount of orange, reddish or blueish-red pigment on the tip of the head; occasionally the entire head may be orange ....*Cephalothrix rufifrons* (p. 54)

71(68). Head forming an obvious rounded cephalic lobe which is wider than the adjacent body regions **72**

Head elongate, pointed or bluntly rounded but not forming a distinct cephalic lobe **74**

72(71). Body more or less uniformly coloured orange-red or reddish-brown, at least on dorsal surface **73**

Head faintly tinged red or translucent and milky-white, intestinal region whitish, often with faint orange-brown or yellow-brown hue; when disturbed the nemerteans secrete large amounts of sticky mucus .... **Tubulanus linearis* (p. 63)

73(72). Body colour a rich reddish- or orange-brown, sometimes more golden-yellow and translucent posteriorly; dorsal and ventral surface equally pigmented ....*Tubulanus polymorphus* (p. 68)

Body orange-red or dark orange dorsally, paler ventrally, with a sharp division between the two sides, or uniformly coloured and with a small white patch on the tip of the head .... **Tubulanus miniatus* (p. 64)

74(71). Head elongate, pointed, body broadest in posterior regions; (*body wall musculature three layered throughout*) **75**

Head bluntly rounded, body broadest in anterior regions; head and anterior body whitish, intestinal region pale buff, tail translucent; (*body wall with two muscle layers extending the full body length, but with additional layers anteriorly in foregut region*) .... *Carinoma armandi* (p. 50)

75(74). Head white or grey-white, most of body pink, yellowish-grey, cinnabar or chocolate brown; (*body wall muscles arranged into outer longitudinal, middle circular and inner longitudinal layers*) .... *Valencinia longirostris* (p. 110)

Head and anterior regions a transparent watery milk white colour, intestinal regions similarly coloured except in sexually mature individuals where the gonads are rosy-brown; head distinctly wrinkled when contracted; (*body wall muscles arranged into outer circular, middle longitudinal and inner circular layers*)................................... **Carinesta anglica* (p. 57)

Systematic part

In the descriptions of species references and synonyms given for each are not intended to be necessarily complete. Apart from those referring to the naming authority, they are based only upon the literature which is specifically concerned with British nemerteans.

M.B.A. in the reference lists refers to the Plymouth Marine Fauna series published by the Marine Biological Association of the United Kingdom.

Anatomical features listed under specific internal characters are either additional to those provided in the generic diagnoses or, where the structure varies between members of the same genus, relate to the condition prevailing in the appropriate form.

Family CARINOMIDAE

Genus *CARINOMA* Oudemans, 1885

Diagnosis

Palaeonemerteans with two principal body wall muscle layers (outer circular, inner longitudinal) reaching the full length of the body, but with additional layers present anteriorly consisting of an outer longitudinal zone between the epidermis and outer circular muscles, a diagonal layer between the outer circular and inner longitudinal musculature, and an inner circular layer adjacent to the foregut and anterior rhynchocoel; anterior epidermis with intra-epithelial muscle fibre network; cerebral ganglia and anterior portions of lateral nerve cords located outside outer circular musculature, but lateral nerves running among inner longitudinal muscle fibres in intestinal regions; nervous system with neurochord cells in cerebral ganglia and neurochord fibres in lateral nerve cords; proboscis with two (outer circular, inner longitudinal) or three (outer and inner circular, middle longitudinal) muscle layers; blood vascular system with six post-cerebral vessels in foregut region, comprising paired main lateral, lateral rhynchocoel and rhynchocoelic villar vessels; intestine with lateral diverticula; eyes and cerebral sensory organs absent, but dorsal surface of head with single median row of sensory ciliated pits; sexes separate.

Carinoma armandi (McIntosh, 1875)
(Fig. 9)

Valencinia armandi McIntosh, 1875a
Carinella armandi McIntosh, 1875b
Carinoma armandi Riches, 1893; Bergendal, 1902, 1903

Specific internal characters

Proboscis with two muscle layers; lateral rhynchocoel and rhynchocoelic villar vessels posteriorly blind-ending.

Description

Although up to about 20 cm long, the body of *Carinoma armandi* (Fig. 9) is only about as thick as a stout thread when extended. The rounded head and anterior regions are whitish in colour, whereas the posterior parts appear pale buff but terminate in a translucent pointed tail. There are 8–12 cephalic epidermal sensory pits in this species.

Carinoma armandi has so far been found at only two localities, both within the British Isles. At the type locality, Southport, large numbers were found in sand near low water level among the tubes of the polychaete *Lanice conchilega* (Pallas) and the nemerteans appear to live in burrows, since in the laboratory individuals readily occupy sandy cases or tubes. Other nemertean species found at the same locality include the heteronemertean *Cerebratulus fuscus* and the hoplonemerteans *Amphiporus hastatus* and *Amphiporus lactifloreus*.

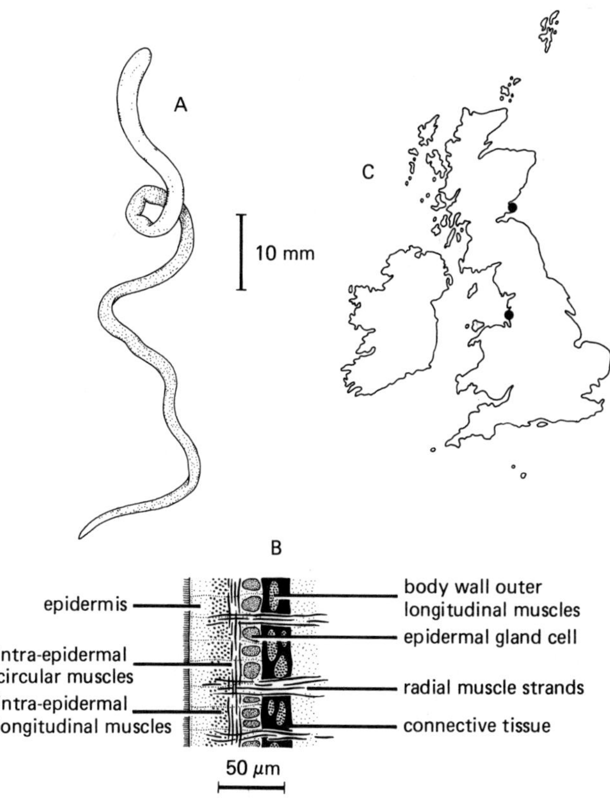

Fig. 9. *Carinoma armandi*; A, dorsal view of whole animal; B, schematic representation of a part of the epidermis and sub-epidermal body wall to show the arrangement of the intra-epithelial muscle fibre network; C, recorded distribution of *Carinoma armandi* in the British Isles.

Family CEPHALOTHRICIDAE

Genus *CEPHALOTHRIX* Örsted, 1843

Diagnosis

Palaeonemerteans with body wall musculature two layered (outer circular, inner longitudinal), without inner circular stratum; cerebral ganglia and lateral nerve cords situated in body wall longitudinal muscle layer; nervous system with neither neurochords nor neurochord cells; proboscis with outer circular and inner longitudinal muscle layers; blood vascular system consisting of paired lateral longitudinal vessels transversely connected only by cephalic and anal lacunae, without mid-dorsal vessel; eyes and cerebral sensory organs absent; mouth located far behind cerebral ganglia; longitudinal muscle plate present between foregut and rhynchocoel; cephalic region with four large nerves; sexes separate.

Cephalothrix linearis (Rathke, 1799)
(Fig. 10A)

Planaria linearis Rathke, 1799
? *Cephalothrix lineata* Claparède, 1862
? *Cephalothrix lineatus* Lankester, 1866
Cephalothrix linearis McIntosh, 1873–74 (in part), 1875c; Riches, 1893; Sheldon, 1896; Gemmill, 1901; M.B.A., 1904, 1931, 1957; Evans, 1909; Elmhirst, 1922; Horsman, 1938; Holme, 1949; Eales, 1952; Crothers, 1966; Gibson & Jennings, 1967; Gibson, 1968a; Jennings & Gibson, 1969; Laverack & Blackler, 1974; Slinger & Gibson, 1974

Specific internal characters

The only obvious internal difference between this species and *Cephalothrix rufifrons* (p. 54) recorded by Wijnhoff (1913) is the absence of parenchymatous connective tissues from the pre-oral regions of *Cephalothrix linearis*.

Description

Cephalothrix linearis possesses a thread-like body, 10–30 cm long but only 0.5–1.0 mm wide. The head is long and bluntly pointed, with the mouth placed far behind the cerebral ganglia; during feeding (under laboratory conditions on small oligochaetes and other annelids) the head is arched dorsally (Fig. 10A) so that the mouth can dilate to ingest the food. The colour is white, creamish or of a pale grey translucent appearance, commonly with a yellowish tinge in the anterior regions. The intestinal tract may be coloured by gut contents.

Cephalic glands are present. The gonads, which may contain mature gametes during the period January to June, are located alongside the lateral blood vessels.

Although *Cephalothrix linearis* is more readily found intertidally, from mid-shore level downwards, it may be dredged from sublittoral habitats. The

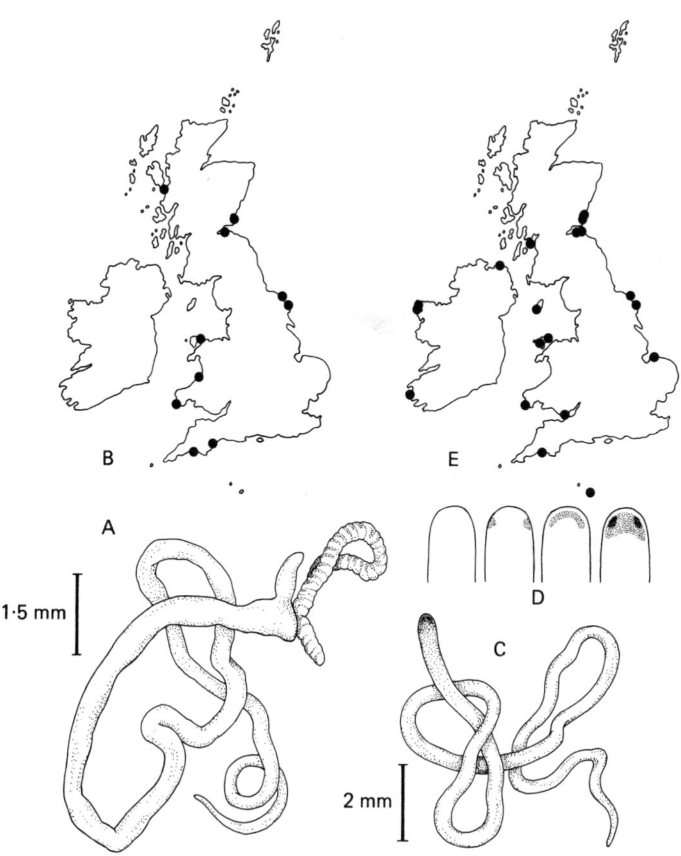

Fig. 10. *Cephalothrix linearis*; A, general view of whole animal about to ingest an oligochaete; note the way in which the head is arched dorsally to allow the oral margins to be distended; B, recorded distribution in the British Isles. *Cephalothrix rufifrons*; C, dorsal view of whole animal; D, variations in the degree of cephalic pigmentation; E, recorded distribution in the British Isles. D redrawn from Hylbom (1957).

species burrows into clean coarse sand, muddy sand or mud to depths of 25–30 cm, but is also found beneath stones and boulders, among the roots of *Laminaria*, or crawling among small algae such as *Corallina* or on hydroid coelenterates like *Tubularia*. When irritated, *Cephalothrix linearis* may contract into a tight knot, but it does not coil spirally like species of *Procephalothrix*, with which it may be confused.

The geographic range of the species extends from the Mediterranean to the west coasts of northern Europe, Greenland, the east coast of North America north of Cape Cod, and to the coasts of Japan.

Cephalothrix rufifrons (Johnston, 1837)
(Fig. 10C, D)

Nemertes Borlasia rufifrons Johnston, 1837
Borlasia rufifrons Johnston, 1846
Gordius gracilis Dalyell, 1853
Astemma rufifrons Johnston, 1865; Lankester, 1866; Parfitt, 1867
Cephalothrix filiformis McIntosh, 1869 (in part)
Polia filum Koehler, 1885
Cephalothrix biocculata Jameson, 1898
Cephalothrix linearis McIntosh, 1873–74 (in part); Stephenson, 1911
Cephalothrix bioculata Riches, 1893; Herdman, 1894, 1900; Vanstone & Beaumont, 1894, 1895; Beaumont, 1895a, b, 1900a, b; Gamble, 1896; Sheldon, 1896; M.B.A., 1904; Gibson & Jennings, 1967; Gibson, 1968a; Jennings & Gibson, 1969
Cephalothrix rufifrons McIntosh, 1867a; Wijnhoff, 1912; Southern, 1913; Evans, 1915; Farran, 1915; M.B.A., 1931, 1957; Smith, 1935; Moore, 1937; Eales, 1952; Barrett & Yonge, 1958; Bruce *et al.*, 1963; Crothers, 1966; Williams, 1972; Khayrallah & Jones, 1975; Campbell, 1976; Boyden *et al.*, 1977

Specific internal characters

Well developed parenchymatous connective tissues in the pre-oral regions.

Description

Anatomically very similar to the preceding species, *Cephalothrix rufifrons* is smaller, attaining a length of 50–60 mm or more but only 0.5 mm or less in width. The colour is translucent or whitish, with a variable amount of orange, red or blueish-red pigment on the tip of the bluntly rounded head (Fig. 10C, D). Occasionally the entire head may exhibit an orange hue.

A gregarious species, *Cephalothrix rufifrons* is commonly found intertidally beneath stones and boulders, usually in clean coarse sand or shelly sand, less often in black muds. It also occurs among algae, particularly on *Corallina* in rock pools. In shallow water it may occur on *Zostera*, but it has also been dredged from sand at depths up to 30 m. The species is occasionally found in conditions of reduced salinity, i.e., the lower regions of estuaries. Breeding takes place from early spring to late autumn. Newly hatched juveniles possess two small black eyes, but these are soon lost and the adults are eyeless.

Cephalothrix rufifrons is known from northern European and Mediterranean coasts and appears to be fairly common, although it is an unobtrusive species.

Genus *PROCEPHALOTHRIX* Wijnhoff, 1913

Diagnosis

Palaeonemerteans with two principal body wall muscle layers (outer circular, inner longitudinal), but with an incompletely developed inner circular stratum in the foregut region; cerebral ganglia and lateral nerve cords situated in body wall longitudinal muscle layer; nervous system with neither neurochords nor neurochord cells; proboscis with outer circular and inner longitudinal muscle layers; blood vascular system consisting of paired lateral longitudinal vessels tranversely connected only by cephalic and anal lacunae, without mid-dorsal vessel; cerebral sensory organs absent; some species with two minute eyes, others eyeless; mouth either far behind cerebral ganglia or not widely separated from them; longitudinal muscle plate present immediately adjacent to dorsal foregut wall; foregut with circular splanchnic muscle layer; cephalic region with four large nerves; sexes separate.

Procephalothrix filiformis (Johnston, 1828–29)
(Fig. 11)

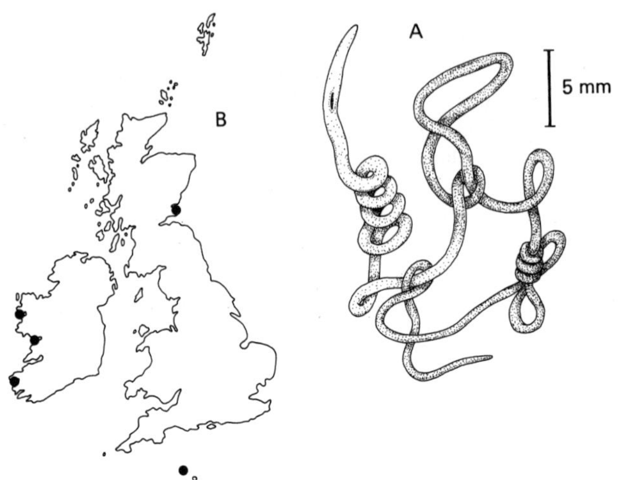

Fig. 11. *Procephalothrix filiformis*; A, general view of whole animal; the head is shown in ventral aspect with the slit-like mouth visible; B, recorded distribution of *Procephalothrix filiformis* in the British Isles.

Planaria filiformis Johnston, 1828–29
Borlasia? *filiformis* Johnston, 1846
Astemma filiformis Johnston, 1865; Lankester, 1866
Cephalothrix filiformis McIntosh, 1867a, 1868a, 1869 (in part)
Procephalothrix filiformis Southern, 1913

Specific internal characters

No nerve layer present between the epidermal basement membrane and the outer body wall circular muscle layer.

Description

Procephalothrix filiformis (Fig. 11) is up to about 15 cm long and 1 mm or more wide. The long bluntly pointed head is generally white or translucent, the remainder of the body a yellowish-white to orange colour which is often darker in the posterior half of the body.

Anatomically very similar to species of *Cephalothrix*, *Procephalothrix filiformis* can be distinguished by the way it contracts in a tight spiral when disturbed. Internally the inner circular muscle fibres associated with the foregut region separate the species from members of the genus *Cephalothrix*.

The species is found beneath stones in muddy situations or buried in muddy gravel from the mid-shore level down to depths of 40 m or more. Its known geographic distribution is restricted to the British Isles and the coast of France. Sexually mature specimens have been found in February.

Family TUBULANIDAE

Genus *CARINESTA* Punnett, 1900

Diagnosis

Palaeonemerteans with body wall musculature comprising outer circular and inner longitudinal layers, sometimes with an inner circular layer; without muscle fibre crosses in the body wall; cerebral ganglia and lateral nerve cords located between epidermal basement membrane and body wall outer circular musculature; epidermal basement membrane thin, usually bordered by thick underlying neural layer in cephalic regions; nervous system with neither neurochord cells nor neurochords; proboscis insertion just behind cerebral ganglia, proboscis with outer circular and inner longitudinal muscle layers; blood vascular system without mid-dorsal vessel; intestine without lateral diverticula; eyes, cerebral and lateral sensory organs absent; sexes separate.

Carinesta anglica Wijnhoff, 1912

Carinesta anglica Wijnhoff, 1912; M.B.A., 1931, 1957

Specific internal characters

Cephalic glands present; inner circular muscle layer missing.

Description

Known only from one specimen obtained from a muddy sandbank at low water level in the River Yealm estuary and one fragment found crawling in sand at Whitsand Bay, near Plymouth, *Carinesta anglica* has never been illustrated and is an inadequately described species. The head is elongate and pointed, becoming obviously wrinkled when contracted. A transparent watery milk-white colour anteriorly, with the proboscis and intestine easily visible through the body wall, the posterior region possesses a rosy-brown tinge due to the gonads. Wijnhoff (1912) commented that the swollen and contracted posterior part of the body fragmented spontaneously when touched.

Genus *TUBULANUS* Renier, 1804

Diagnosis

Palaeonemerteans with body wall musculature consisting of outer circular, middle longitudinal and inner circular layers, often with dorsal and/or ventral muscle fibre crosses passing between the two circular strata; cerebral ganglia and lateral nerve cords situated between epidermal basement membrane and body wall outer circular musculature; epidermal basement membrane usually thick; nervous system with neither neurochords nor neurochord cells; proboscis with outer circular and inner longitudinal muscle layers; blood vascular system without mid-dorsal vessel; intestine with shallow lateral diverticula; eyes absent; cerebral sensory organs consisting of simple ciliated canals in the cephalic epidermis; some species with lateral sensory organs in foregut region; sexes separate.

Tubulanus albocapitatus Wijnhoff, 1912
(Fig. 12A)

Tubulanus albocapitatus Wijnhoff, 1912; M.B.A., 1931, 1957

Specific internal characters

Not known; no information has ever been published concerning the species' internal anatomy.

Description

Tubulanus albocapitatus (Fig. 12A) is reported as 15 mm long and 0.5 mm wide in the foregut region. It does not possess lateral sensory organs. The head, unlike that of many tubulanids, is neither sharply demarcated from the body nor developed into a distinctive cephalic lobe.

Wijnhoff (1912: 412–13), whose is the only description of the species, gives the following account of the colour pattern: 'The perfectly white head is followed by a brown-red belt of the same breadth, which is the darkest part of the whole body. A yellow pigment is distributed all over this region . . . The first circular white line separates this region from the body. The median dorsal line, which is white too, passes through this belt into the dark region described above, but does not reach the white head. At both sides of this median longitudinal line, separated from it by a translucent region, a reddish-brown stripe of a fainter tint is present. These stripes do not even reach the lateral lines, as a transparent region is developed between them too; the white pigment patches, which are dispersed at the sides, constitute a very inconspicuous and incomplete lateral line. The ventral side is also transparent. Eleven white belts are present . . . Just in front of the second ring one pair of white patches is conspicuous; they are connected with this belt and are situated in the dorsal red stripes.'

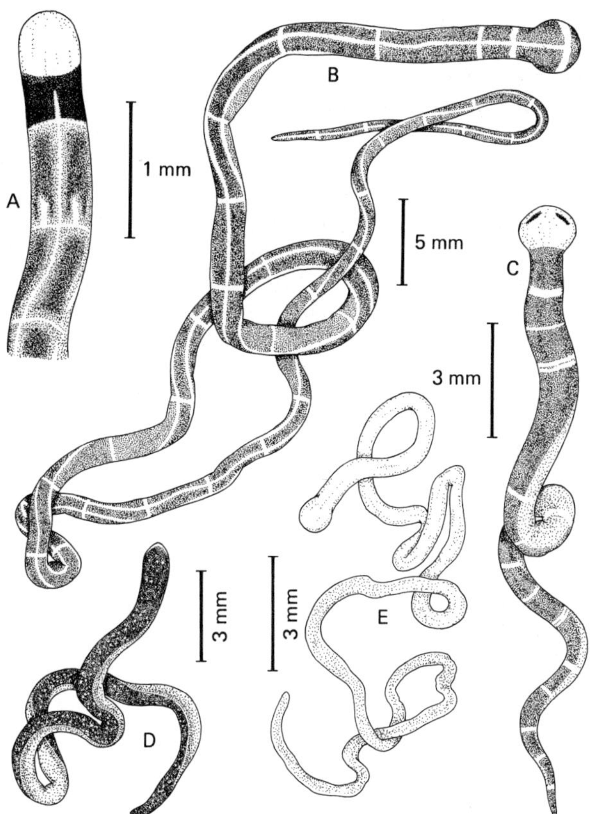

Fig. 12. *Tubulanus albocapitatus*; A, dorsal view of anterior region. *Tubulanus annulatus*; B, general view of whole animal. *Tubulanus banyulensis*; C, dorsal view of whole animal. *Tubulanus inexpectatus*; D, general view of whole animal. *Tubulanus linearis*; E, general view of whole animal. A redrawn from and based on description given by Wijnhoff (1912), C, E redrawn from Bürger (1895), D based on description given by Beaumont (1900b).

This species, which is inadequately described, is known only from three specimens dredged on different occasions from the Rame-Eddystone grounds near Plymouth (Fig. 14A). The grounds have depths of 50–60 m and are typically either muddy gravel or fine muddy sand.

Tubulanus annulatus (Montagu, 1804)
(Fig. 12B)

Gordius annulatus Montagu, 1802, 1804, 1808 (in part); Turton, 1807; Pennant, 1812; Thompson, 1841
Lineus annulatus Montagu, 1808
Carinella trilineata Johnston, 1833; Thompson, 1841
Meckelia trilineata Thompson, 1843, 1856; Johnston, 1846
Meckelia annulata Johnston, 1865 (in part); Lankester, 1866; Parfitt, 1867; McIntosh, 1868b, 1869; Chumley, 1918
Carinella annulata McIntosh, 1873–74 (in part), 1875a, b, c, 1927; Haddon, 1886a, b; Riches, 1893; Herdman, 1894; Beaumont, 1895a, b, 1900a, b; Vanstone & Beaumont, 1894, 1895; Gamble, 1896; Sheldon, 1896; Jameson, 1898; Allen, 1899; Allen & Todd, 1900; Gemmill, 1901; M.B.A., 1904; Evans, 1909; Horsman, 1938
Carinella linearis Gibson, 1886
Carinella mcintoshii Riches, 1893
Carinella macintoshi Sheldon, 1896
Carinella aragoi Beaumont, 1895a, b; Gamble, 1896; Herdman, 1900
Tubulanus annulatus Southern, 1908a, 1913; Stephenson, 1911; Massy, 1912; Wijnhoff, 1912; Farran, 1915; M.B.A., 1931, 1957; Moore, 1937; Jones, 1951; Eales, 1952; Bassindale & Barrett, 1957; Barrett & Yonge, 1958; Bruce *et al.*, 1963; Crothers, 1966; Laverack & Blackler, 1974; Campbell, 1976; Boyden *et al.*, 1977

Specific internal characters

Body wall inner circular muscle layer well developed; no dorsal body wall muscle cross; cephalic glands well developed; lateral sensory organs absent.

Description

One of the most strikingly coloured of the British nemertean species, *Tubulanus annulatus* is a vivid brick-red, orange-red, garnet-red or brownish-red marked with white longitudinal stripes and rings (Fig. 12B). The mid-dorsal stripe anteriorly extends on to the distinctive rounded cephalic lobe, which is the same colour as the remainder of the body, and terminates at a transverse white head band bordering the pigmented tip of the snout. The two lateral stripes, in contrast, do not reach the head and run posteriorly from the first white ring encircling the body. Apart from the first two or three, the rings are arranged more or less equidistantly. The ventral surface is paler than the dorsal.

The species attains a length of 75 cm or more but rarely exceeds 3–4 mm width unless it is strongly contracted. Although sometimes found intertidally beneath stones, among *Laminaria* roots or on sand or mud near low water level, *Tubulanus annulatus* is more common sublittorally on a wide variety of substrata (gravel, stones, mud, fine sand, sand mixed with shell fragments, among scallops) at depths up to 40 m or more. It readily secretes a silken mucous tube to which sediment particles adhere. In the British Isles (Fig. 14B) *Tubulanus annulatus* reproduces during the months of June and July.

The species possesses a wide geographic range in the northern hemisphere, ranging from the Pacific coast of North America eastwards to the Atlantic, North Sea and Mediterranean coasts of Europe. A record of the species from South Africa (Stimpson, 1857) is of dubious validity. The internal morphology of *Tubulanus annulatus* apparently varies with its geographic distribution and is discussed by Hylbom (1957: 571–4).

Tubulanus banyulensis (Joubin, 1890)
(Fig. 12C)

Carinella banyulensis Joubin, 1890
Tubulanus banyulensis Southern, 1913; Farran, 1915

Specific internal characters

Lateral sensory organs present but poorly developed.

Description

Tubulanus banyulensis is an inadequately described species. A small form, 10–15 mm long and 1.0–1.5 mm wide, it possesses a cylindrical body which posteriorly narrows to a bluntly pointed tail. The head, which is anteriorly rounded but only a little wider than the body, is white or pinkish and bears two small patches of black pigment near its anterior margins (Fig. 12C). According to Bürger (1895) these patches are dorsal, but Southern (1913) observes that they occur on the underside of the head. The dorsal body colour varies from a bright cherry-red or reddish-brown to a muddy green, the ventral surface is pinkish or yellowish; there is a sharp division between the upper and lower colours and some examples have been reported with a faint red or white lateral longitudinal stripe on each side of the body. An indistinct mid-dorsal stripe, occasionally running into the cephalic pigment, may appear as a longitudinal series of white flecks or be entirely absent. Up to about 18 white transverse rings encircle the body; these may be more or less equally spaced or be loosely grouped in threes with the first and third ring of each group wider than the middle one and sometimes appearing double.

In the British Isles *Tubulanus banyulensis* has been found on only four occasions, dredged from 4–16 m depth off the north-western coast of Ireland (Fig. 14A). The species is also recorded from the Mediterranean.

Tubulanus inexpectatus (Hubrecht, 1880)
(Fig. 12D)

Carinella inexpectata Hubrecht, 1880; Beaumont, 1900a, b
Tubulanus inexpectatus Southern, 1913

Specific internal characters

Ciliated canal of cerebral sensory organs penetrates cerebral ganglia.

Description

An insufficiently described species, *Tubulanus inexpectatus* (Fig. 12D) is up to 35 mm long and 1 mm wide, of an overall brilliant vermilion or reddish-brown colour apart from the anterior portion of the lancet-shaped head, which is colourless. The ventral surface is rather paler than the dorsal. Minute specks of opaque white are scattered over the body surface, and along each side of the body run pale lines which near the head curve ventrally to meet just anterior to the mouth.

Only a single specimen of this species has been found in the British Isles. It was dredged from about 90 m depth south-west of Bray Head, Valencia Island, south-western Ireland (Fig. 14A), on clean worn gravel. *Tubulanus inexpectatus* has been recorded from only one other locality, Capri in the Mediterranean.

Tubulanus linearis (McIntosh, 1873–74)
(Fig. 12E)

Carinella linearis McIntosh, 1873–74, 1875a, b; Riches, 1893; Sheldon, 1896; Benham, 1897; M.B.A., 1904
Tubulanus linearis Wijnhoff, 1912; Southern, 1913; Farran, 1915; M.B.A., 1931, 1957

Specific internal characters

Body wall with dorsal and ventral muscle crosses; lateral sensory organs present; cephalic glands absent.

Description

Tubulanus linearis is a poorly described form. It is a moderately long (up to about 15 cm) but slender (0.5–1.0 mm) species with a flattened and rather spatulate head and tapering body (Fig. 12E). The nemerteans produce copious amounts of a viscid mucus and readily form delicate tubes to which sand grains adhere.

Most specimens are an overall pure milky-white, although the intestinal regions may be faintly tinged orange-brown or yellowish-brown. The posterior body margins are often translucent. In the head, which may also appear translucent or exhibit a faint reddish tint, the rhynchodaeum is sometimes distinguishable as a median white line in front of the cerebral ganglia.

A lower-littoral to shallow water species, *Tubulanus linearis* is found burrowed in clean sand or among sand-binding algae, sometimes associated with polychaetes of the genus *Magelona*. Apart from the British Isles (Fig. 14C), the species has also been found on the French and Italian coasts.

Tubulanus miniatus (Bürger, 1892)
(Fig. 13A)

Carinella miniata Bürger, 1892
Tubulanus miniatus Wijnhoff, 1912; M.B.A., 1931, 1957

Specific internal characters

Not known; no information is available on the internal morphology of this species.

Description

Originally and inadequately described by Bürger (1892) from a solitary specimen dredged from 70 m depth near Naples, Wijnhoff (1912) regarded three individuals obtained from a depth of 45–55 m on the Rame-Eddystone grounds near Plymouth (Fig. 14A) as belonging to this species. Differences in colour pattern described by these two authors, however, suggest that the British record is possibly of a different species and thus of questionable validity. Bürger's nemertean, 4.5 cm long and 2 mm wide, was uniformly orange-red in colour (Fig. 13A), whereas Wijnhoff's examples, the largest 3 cm long, were dark orange with the dorsal surface more intensely pigmented than the ventral and the two regions sharply demarcated by a line. Wijnhoff also mentions a white patch on the tip of the head which is not described by Bürger.

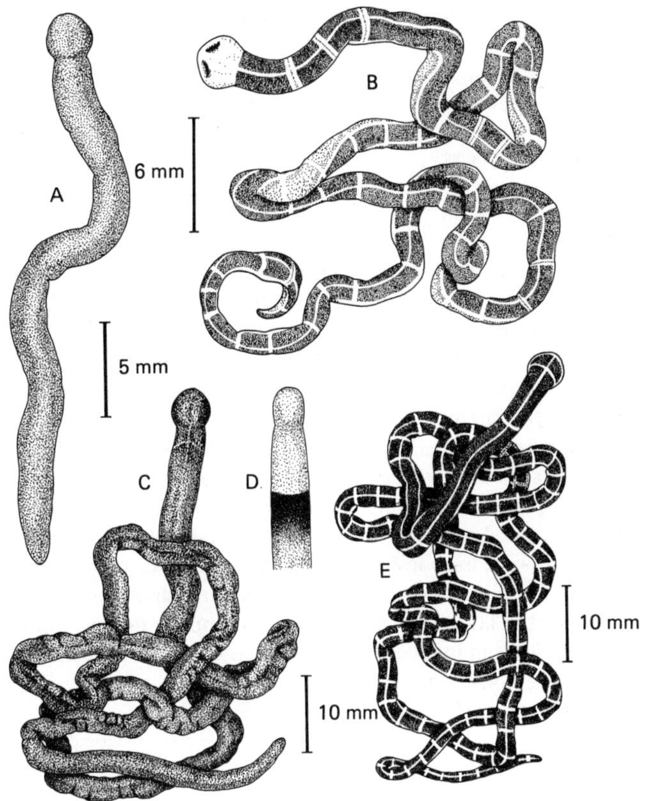

Fig. 13. *Tubulanus miniatus*; A, dorsal view of whole animal. *Tubulanus nothus*; B, general view of complete specimen. *Tubulanus polymorphus*; C, general view of whole animal; D, dorsal view of anterior end to show the characteristic pattern of pigmentation after preservation. *Tubulanus superbus*; E, general view of complete specimen. A, B redrawn from Bürger (1895).

Tubulanus nothus (Bürger, 1892)
(Fig. 13B)

Carinella nothus Bürger, 1892
Tubulanus nothus Wijnhoff, 1912; M.B.A., 1931, 1957

Specific internal characters

Lateral sensory organs present.

Description

Up to 10 cm long, the body is anteriorly cylindrical, posteriorly broad and flattened; Bürger in both 1892 and 1895 records the width as 2.0–2.5 cm but, from his 1895 illustration (Plate 1, Fig. 12) this should be 2.0–2.5 mm. The colourless head, marked by a pair of black crescent-shaped pigment patches near its anterior margins (Fig. 13B), may be up to twice the width of the body depending upon the degree of contraction. The dorsal body surface is a dark dirty reddish-brown colour, the ventral surface a paler orange-brown fading to yellowish towards the two extremities. White dorsal median and lateral longitudinal stripes are present, and there is usually an inconspicuous mid-ventral line too. Forty or more white rings encircle the body, the first three located some distance in front of the remainder. Occasional rings may appear double. The lateral sensory organs appear as small orange pits on the dorsal margin of the lateral stripes, just in front of the fourth white ring.

Tubulanus nothus, an inadequately described species, is known only from the Plymouth (Fig. 14A) and Naples areas.

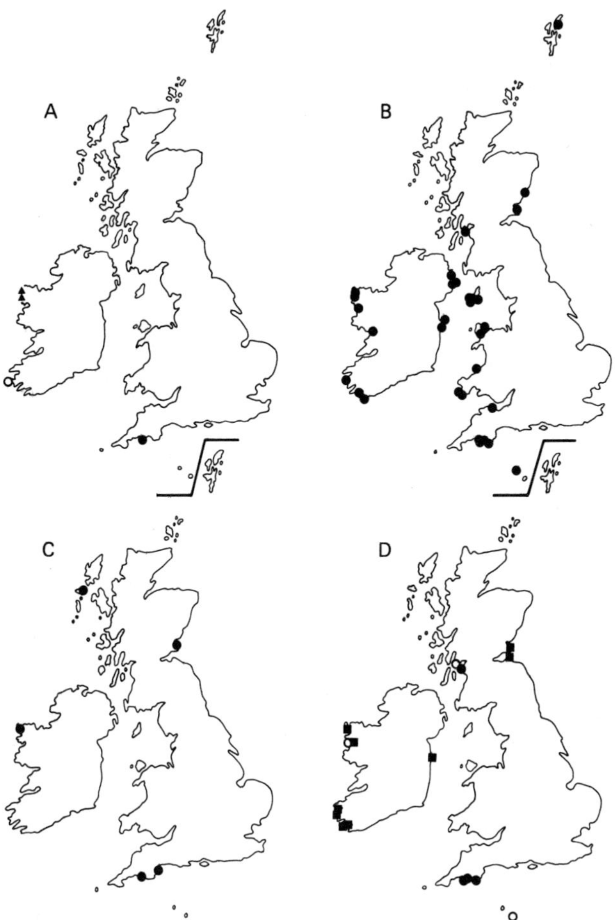

Fig. 14. A, recorded distribution of *Tubulanus albocapitatus*, *Tubulanus miniatus*, *Tubulanus nothus* (●), *Tubulanus inexpectatus* (○) and *Tubulanus banyulensis* (▲) in the British Isles; B, recorded distribution of *Tubulanus annulatus* in the British Isles; C, recorded distribution of *Tubulanus linearis* in the British Isles; D, recorded distribution of *Tubulanus polymorphus* (○), *Tubulanus superbus* (■) or both species together (●) in the British Isles.

Tubulanus polymorphus Renier, 1804
(Fig. 13C, D)

Tubulanus polymorphus Renier, 1804; King, 1911; Wijnhoff, 1912; Southern, 1913; M.B.A., 1931, 1957
Valencia splendida Koehler, 1885
Carinella polymorpha Riches, 1893; Sheldon, 1896; Allen & Todd, 1900; M.B.A., 1904

Specific internal characters

Body wall with dorsal muscle cross only; lateral sensory organs present.

Description

Up to 50 cm or more long and 0.5 cm wide, *Tubulanus polymorphus* (Fig. 13C, D) possesses an extremely soft, pliable body uniformly coloured a beautiful rich reddish- or orange-brown. The cephalic lobe is rounded and distinct, much broader than the trunk. In the posterior intestinal regions the body may appear somewhat translucent and golden-yellowish in hue.

Externally *Tubulanus polymorphus* closely resembles a Scandinavian species, *Tubulanus théeli* (Bergendal, 1902); *Tubulanus théeli*, however, possesses both dorsal and ventral muscle crosses between the body wall circular muscle layers (Hylbom, 1957). In alcohol or formalin preserved specimens of *Tubulanus polymorphus* the general colour changes to a dull creamy-white marked by a characteristic band of dark reddish-brown encircling the body in the foregut region (Fig. 13D). The width of this pigmented band is variable, but its anterior margin is invariably precisely marked, in contrast to the posterior limits which may be sharply defined also or merely fade to the remaining body colour.

Tubulanus polymorphus has a wide geographic range in the northern hemisphere, extending from the Pacific coast of North America eastwards to the Mediterranean and northern coasts of Europe. In the British Isles (Fig. 14D) it is found on sandy or gravelly bottoms at depths up to 45–50 m or more.

Tubulanus superbus (Kölliker, 1845)
(Fig. 13E)

Gordius annulatus Montagu, 1808 (in part)
Nemertes superbus Kölliker, 1845
Gordius anguis Dalyell, 1853
Meckelia annulata Johnston, 1865 (in part)
Meckelia annulata var. Parfitt, 1867
Carinella annulata McIntosh, 1873–74 (in part)
Carinella superba Jameson, 1898; Allen, 1899; Allen & Todd, 1900; Beaumont, 1900a, b; Punnett, 1901a; M.B.A., 1904; Elmhirst, 1922
Tubulanus superbus Southern, 1908a, 1913; Wijnhoff, 1912; Farran, 1915; M.B.A., 1931, 1957

Specific internal characters

Body wall inner circular muscle layer well developed; lateral sensory organs present; cephalic glands absent.

Description

One of the largest British nemertean species, *Tubulanus superbus* (Fig. 13E) may attain a length of over 75 cm and width of 5 mm. The body is firm in consistency, gradually narrowing behind the distinct rounded head to a bluntly pointed tail. The colour is a rich dark brown, reddish-brown, scarlet or cherry-red, marked with mid-ventral, mid-dorsal and lateral longitudinal white or yellowish-gold stripes and large numbers (up to 200 or more) of white rings. The first two or three rings are fairly widely spaced, whereas the remainder are closer together and more or less uniformly distributed along the body. Occasional rings may appear double.

In many ways resembling *Tubulanus annulatus* (p. 60), these two species may be easily separated by the presence (*Tubulanus superbus*) or absence (*Tubulanus annulatus*) of the distinct median ventral longitudinal stripe.

Tubulanus superbus secretes a mucoid sheath to which sand and other particles adhere. Sometimes found intertidally, the species is much more commonly obtained by dredging on sandy or gravelly sediments at depths of up to 80 m or more. Around the British Isles it spawns during June and July.

The geographic range of *Tubulanus superbus* extends from the Mediterranean to the Channel coasts of France, the British Isles (Fig. 14D) and northwards to Scandinavia.

Family BASEODISCIDAE

Genus *BASEODISCUS* Diesing, 1850

Diagnosis

Heteronemertea without horizontal lateral cephalic slits, some species with shallow oblique grooves; proboscis with two (outer longitudinal, inner circular) muscle layers and no muscle crosses; rhynchocoel short, rarely exceeding one-third of body length; rhynchocoel wall circular musculature not interwoven with body wall muscle layers; dorsal fibrous core of cerebral ganglia branched only at rear into dorsal and ventral points; nervous system with neither neurochords nor neurochord cells; foregut with distinct sub-epithelial gland cell zone, usually at least in part separated from epithelium by longitudinal, circular and/or oblique splanchnic muscle fibres; dermis well developed, with separate outer glandular and inner connective tissue regions; caudal cirrus absent; cephalic glands extensively developed, reaching post-orally among the outer body wall longitudinal muscle fibres; single frontal organ present; excretory system mostly with external nephridiopores, but some species with efferent ducts discharging into foregut; eyes present, minute and numerous; sexes separate.

Baseodiscus delineatus (Delle Chiaje, 1825)
(Fig. 15)

Polia delineata Delle Chiaje, 1825
Eupolia curta Allen, 1899; M.B.A., 1904
Baseodiscus curtus M.B.A., 1931, 1957

Specific internal characters

Splanchnic musculature composed of mixed longitudinal and oblique fibres; excretory system without internally opening efferent tubules.

Description

Two species of *Baseodiscus*, *Baseodiscus delineatus* and *Baseodiscus curtus* (Hubrecht, 1879), exhibit a 'remarkable (morphological) similarity to each other' (Gibson, 1979: 146) and, indeed, *Baseodiscus curtus* has been regarded merely as a variety of *Baseodiscus delineatus* by such authors as Coe (1940, 1944) and Corrêa (1958, 1961, 1963). Since the specific name *delineatus* has priority, and because there are strong grounds for considering the two forms as conspecific, the British species is regarded as *Baseodiscus delineatus*.

The species may attain lengths of 1 m or more but is rarely more than 2–3 mm wide except when fully contracted. The slightly bilobed head (Fig. 15B) is clearly distinguishable from the remainder of the body and may be almost completely retracted into the trunk when the animal contracts strongly. There are numerous small black or dark brown eyes; these are mainly distributed along the cephalic margins but near the back of the head

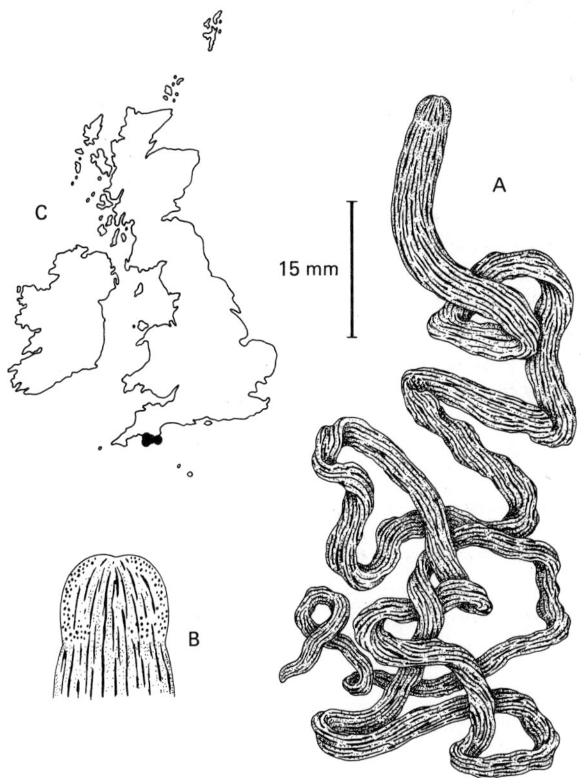

Fig. 15. *Baseodiscus delineatus*; A, general view of whole animal; B, dorsal view of cephalic region to show the distribution of the eyes; C, recorded distribution of *Baseodiscus delineatus* from the British Isles.

extend medially to form two large dorsolateral groups. The large mouth is obvious on the ventral surface just behind the head.

The background colour is a uniform dull yellowish-fawn to light brown, marked by reddish-brown interrupted longitudinal stripes which extend the full body length. The stripes are very variable in width and outline and adjacent stripes may fuse with each other. The number of stripes at any part of the body ranges from about 5–12 on the dorsal surface; generally there are rather fewer stripes on the ventral margins and these are often paler than their dorsal counterparts.

Baseodiscus delineatus is one of the most widespread nemertean species and possesses a circumglobal distribution spanning both hemispheres. Mostly known from tropical and subtropical localities (Gibson, 1979), the species appears to reach its northernmost limits in British waters, where it has only been found sublittorally on coarse shell or gravel substrata containing some sand or mud.

Genus *OXYPOLIA* Punnett, 1901

Diagnosis

Heteronemertea without lateral horizontal cephalic slits but with a transverse furrow encircling the back of the head; proboscis with three muscle layers (outer longitudinal, middle circular, inner longitudinal), without muscle crosses; rhynchocoel wall circular muscles not interwoven with body wall musculature; proboscis pore subterminal, ventral; dorsal fibrous core of cerebral ganglia branched only at rear into dorsal and ventral points: nervous system with two median dorsal nerves; dermal gland cells and connective tissues intermingled, not in distinct zones; caudal cirrus absent; cephalic glands very well developed, posteriorly extending behind cerebral ganglia; frontal organ absent; eyes absent; cerebral sensory organs small, not surrounded by blood lacunae; sexes separate.

Oxypolia beaumontiana Punnett, 1901
(Fig. 16)

Oxypolia beaumontiana Punnett, 1901a; M.B.A., 1904, 1931, 1957

Specific internal characters

The genus *Oxypolia* is monospecific; the specific internal characters are thus as given in the generic diagnosis.

Description

Found on just four occasions, all within the Plymouth area, *Oxypolia beaumontiana* (Fig. 16) is at present known only from the British Isles. The largest specimen obtained was about 12 cm long and 5 mm wide. The body tends to become noticeably flattened when extended, but is otherwise cylindrical in the anterior half apart from the flat and pointed head. The cephalic furrow encircles the back of the head, just in front of the mouth. In the anterior half, including the head, the colour is milk white, in the intestinal regions there is a brownish or pale rose-red tinge.

Oxypolia beaumontiana is found intertidally and to depths of about 30–50 m. Sublittorally it is associated with soft rocks free from algal growth but well covered with colonial invertebrates such as polyzoans, hydroid coelenterates and sponges.

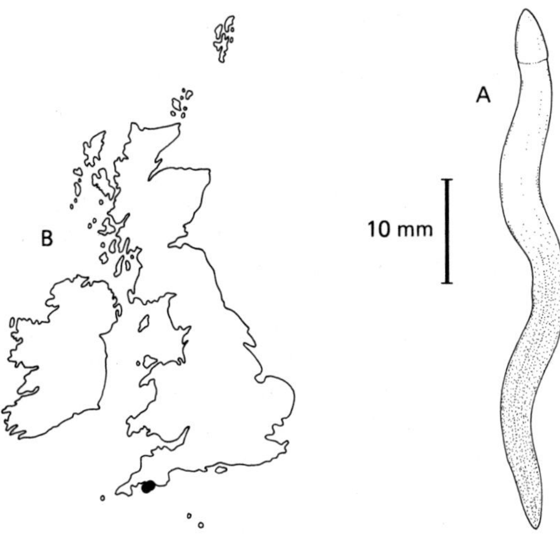

Fig. 16. *Oxypolia beaumontiana*; A, dorsal view of whole animal; B, recorded distribution from the British Isles. A redrawn from Punnett (1901a).

Family LINEIDAE

Genus *CEREBRATULUS* Renier, 1804

Diagnosis

Heteronemertea with a single pair of horizontal lateral cephalic slits; proboscis containing two (outer longitudinal, inner circular) or three (outer longitudinal, middle circular, inner longitudinal) muscle layers and 0–2 muscle crosses; rhynchocoel wall circular musculature not interwoven with body wall muscle layers; dorsal fibrous core of cerebral ganglia branched only at rear into dorsal and ventral points; nervous system with neurochords and neurochord cells; foregut with longitudinal splanchnic muscles; dermis variable, either comprising distinct outer glandular and inner connective tissue layers or with gland cell region abutting directly against body wall outer longitudinal muscles without intervening connective tissues; body wall musculature usually with a diagonal layer; caudal cirrus present; dorsoventral muscles, especially in intestinal regions, usually powerfully developed; cephalic glands usually well developed; most species with frontal organ; eyes present or absent; sexes separate; generally large animals, often with sharpened lateral margins and able to swim actively with sinuous up and down movements.

Cerebratulus alleni Wijnhoff, 1912

Cerebratulus alleni Wijnhoff, 1912; M.B.A., 1931, 1957

Specific internal characters

Body wall without diagonal muscle layer; dermis without distinct connective tissue layer.

Description

Cerebratulus alleni is not a well described species and has never been illustrated. The living worm is recorded as having a light flesh colour with a white cephalic tip. After preservation most of the body is a yellowish-grey tinge, sharply demarcated from the greyish-brown colouration of the anterior end. No dimensions have been reported for the species. The main external features described for *Cerebratulus alleni* are a conspicuously swollen and pear- or fig-shaped head and long but shallow cephalic slits which posteriorly reach almost to the level of the mouth.

Cerebratulus alleni is known from only a single specimen obtained from a sand bank on the River Yealm, near Plymouth (Fig. 18A).

Cerebratulus fuscus (McIntosh, 1873–74)
(Fig. 17A)

Micrura sp. McIntosh, 1869
Micrura fusca McIntosh, 1873–74, 1875a, c; Gamble, 1896
Cerebratulus fuscus Riches, 1893; Sheldon, 1896; Beaumont, 1895a, b, 1900a, b; Herdman, 1896, 1900; Jameson, 1898; Allen, 1899; M.B.A., 1904, 1931, 1957; Wijnhoff, 1912; Southern, 1913; Farran, 1915; Moore, 1937; Bruce, 1948; Bruce *et al.*, 1963; Eagle, 1973; Laverack & Blackler, 1974; Campbell, 1976

Specific internal characters

Dermis with insignificant amounts of connective tissue which do not separate glandular region from body wall muscles; proboscis with three muscle layers, the outer longitudinal zone peripherally incomplete, with one muscle cross.

Description

Cerebratulus fuscus (Fig. 17A) attains a length of up to 15 cm and is 2–4 mm or more wide. The body is characteristically dorsoventrally flattened with thin translucent or opaque margins. Anteriorly the head tapers to a bluntly rounded point, posteriorly the body is either of a more or less uniform width or gradually broadens throughout its length, terminating bluntly and bearing a short but distinct colourless caudal cirrus. There are 4–13 eyes on each side of the head; McIntosh (1873–74) notes that the eyes are black, whereas Cantell (1975) observes that in Swedish examples they are red. Wijnhoff (1912) erroneously reported that the species lacks eyes.

The general colour is a pale yellowish, greyish-brown or pinkish tinge, dorsally speckled lightly and indistinctly with red, brown or greenish-grey. Often the anterior dorsal surface is more darkly pigmented than the rest of the body, although in larger individuals this colouration tends to spread throughout the animal's length. There is usually a distinct reddish hue in the cerebral ganglionic region, and the lateral nerve cords may appear as pinkish longitudinal lines in the intestinal regions. The cephalic slits too may be internally tinged pink or red. The body is commonly marked by pale transverse lines or furrows.

Cerebratulus fuscus may swim actively for a few minutes if disturbed. Occasionally found intertidally, this species is more commonly obtained by dredging at depths to about 50 m but has been found (off Portugal) as deep as 1590 m. Typically living in sandy sediments or amongst shelly gravel, it may sometimes be obtained from laminarian holdfasts, with oyster shells or in mud. The geographic distribution extends from Alaska and Florida in North America to Greenland, European waters (including the Mediterranean) and South Africa (for British distribution see Fig. 18B).

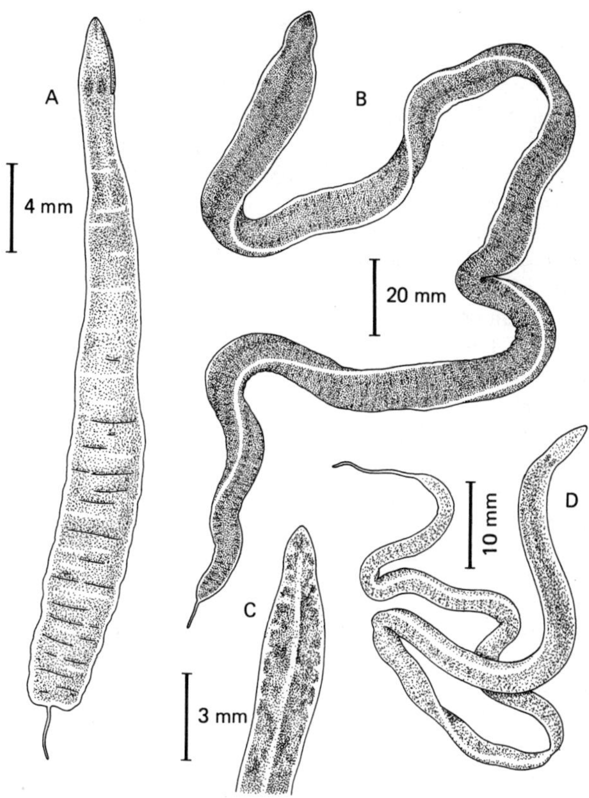

Fig. 17. *Cerebratulus fuscus*; A, dorsal view of complete specimen. *Cerebratulus marginatus*; B, general view of whole animal. *Cerebratulus pantherinus*; C, dorsal view of anterior body regions. *Cerebratulus roseus*; D, general view of whole animal. A, C, D redrawn from Bürger (1895).

Cerebratulus marginatus Renier, 1804
(Fig. 17B)

Cerebratulus marginatus Renier, 1804; Jameson, 1898; Massy, 1912; Southern, 1913; Bassindale & Barrett, 1957; Bruce *et al.*, 1963; Crothers, 1966
Serpentaria fragilis Goodsir, 1845; Johnston, 1846, 1865
Gordius fragilis Dalyell, 1853
Lineus beattiaei Gray, 1857, 1858
Serpentaria beattiei Johnston, 1865
Cerebratulus angulatus McIntosh, 1873–74, 1875b, 1906a, 1927; Haddon, 1886b; Herdman, 1894; Scott, 1894; Vanstone & Beaumont, 1894, 1895; Beaumont, 1895a, b; Sheldon, 1896; Gemmill, 1901; Evans, 1909; Laverack & Blackler, 1974
Cerebratulus angulosus Haddon, 1886b
Cerebratulus lacteus Clark & Cowey, 1958

Specific internal characters

Dermal gland cells not separated from body wall musculature by connective tissue layer; proboscis with three muscle layers and two muscle crosses.

Description

A bulky species, *Cerebratulus marginatus* (Fig. 17B) may reach a length of 1 m and width of 25 mm or more. The worms, however, are exceptionally contractile and may be less than half their extended length. The head is sharply tapered and bluntly pointed, with wide cephalic furrows. Although most authorities state that eyes are absent from this species, Cantell (1975) notes that pigmented eyes are present; their small size, however, makes them easily overlooked in living individuals. The body behind the mouth is dorsoventrally compressed, with sharp lateral margins. Transverse epidermal folds of varying sizes often give the nemerteans an irregularly wrinkled appearance, particularly when the body is contracted. The transparent caudal cirrus is slender.

In colour *Cerebratulus marginatus* is typically a uniform greyish-brown with whitish or transparent lateral margins through which the pink coloured lateral nerve cords are often visible. Younger individuals are generally paler, and the dorsal pigmentation of older animals may be somewhat darker than that of the underside. Variations in colour reported include dark greyish-green, slate-blue or dull brown. Both the mouth and cephalic slits are pale on their inner margins. In young specimens the cerebral ganglia and proboscis apparatus may show through the epidermis.

Rarely found on the lower shore, *Cerebratulus marginatus* is usually obtained when dredging in sandy or muddy sediments at depths of 20–150 m or more, although it may be caught on fishing lines when mussel flesh is used as bait and one large example was found in the stomach of a cod (*Gadus morhua* L.). *Cerebratulus marginatus* is able to swim actively with strong dorsoventral undulatory movements. When swimming it often rotates about

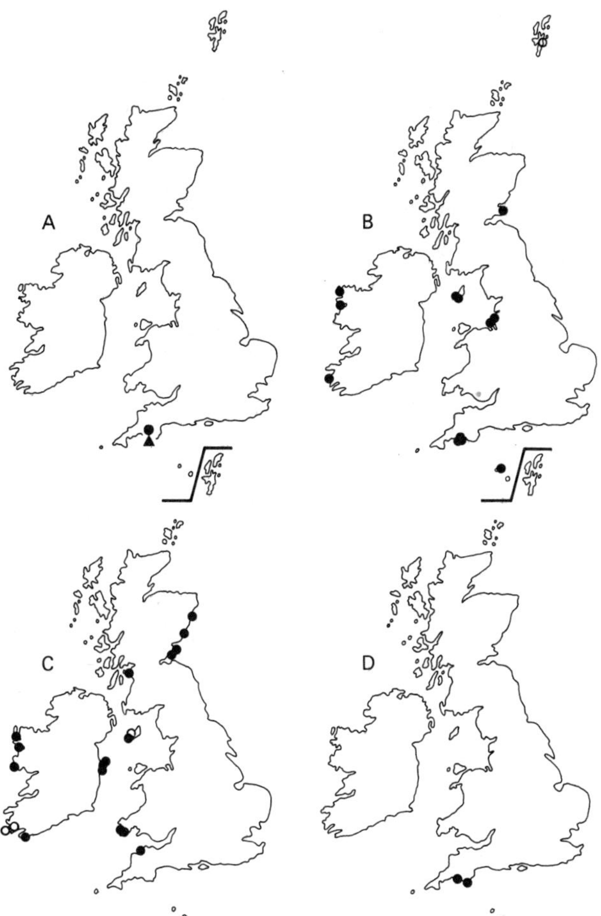

Fig. 18. A, recorded distribution of *Cerebratulus alleni* (●) and *Cerebratulus pantherinus* (▲) from the British Isles; B, recorded distribution of *Cerebratulus fuscus* from the British Isles; C, recorded distribution of *Cerebratulus marginatus* from the British Isles; ● = certain record, ○ = specific identity not confirmed; D, recorded distribution of *Cerebratulus roseus* from the British Isles.

its own longitudinal axis and frequently projects its head above the water surface. It also readily fragments when handled.

The species has a wide geographic range in the northern hemisphere. In European waters, apart from the British Isles (Fig. 18C), it extends from Scandinavia to the Mediterranean and southwards to Madeira; it is widespread in the Arctic (King Charles Land, Bremer Sound, Hinlopen Strait, east Spitzbergen), and is found in the western North Altantic from Greenland, Labrador and Cape Cod southwards beneath the off-shore Arctic current, and in the Pacific from Alaska to San Diego, California on the eastern seaboard and Japan in the west.

Cerebratulus pantherinus Hubrecht, 1879
(Fig. 17C)

Cerebratulus pantherinus Hubrecht, 1879; Riches, 1893; Sheldon, 1896; M.B.A., 1904, 1931, 1957

Specific internal characters

Body wall without a diagonal muscle layer; dermis with separate glandular and connective tissue layers.

Description

A poorly described species, *Cerebratulus pantherinus* (Fig. 17C) has been reported with lengths of 4–7 cm or more. Possessing a typically cerebratulid shape, this species is characteristically marked by a mottled pattern of irregular dirty-green, brownish, yellowish and white pigment patches which are especially evident on the anterodorsal margins. Internally it apparently differs from *Cerebratulus marginatus* (at one time it was regarded as a colour variety of *Cerebratulus marginatus* but was re-established as a separate species by Bürger, 1895) in the shape of the cerebral ganglia and in lacking eyes.

Cerebratulus pantherinus is a deep-water form, dredged from sandy sediments at 50 m or more depth. Only a single British record exists, found off Stoke Point, near Plymouth (Fig. 18A) in 1892. Apart from the British record it has been found only at Roscoff (English Channel) and in the Mediterranean at Naples.

Cerebratulus roseus (Delle Chiaje, 1841)
(Fig. 17D)

Polia rosea Delle Chiaje, 1841
Cerebratulus roseus Wijnhoff, 1912; M.B.A., 1931, 1957

Specific internal characters

Dermis without distinct connective tissue layer; cephalic glands poorly developed.

Description

Cerebratulus roseus (Fig. 17D) is a long (up to about 50 cm) but comparatively slender and inadequately described species, at most only 5–6 mm wide. It possesses an unusually long and thin caudal cirrus which may extend 2 cm beyond the posterior tip of the body. The yellowish coloured head is narrower than the trunk, sharply pointed, and bears long cephalic slits but no eyes. At the rear of the head the cerebral ganglia appear distinctly reddish through the body wall. In the foregut regions the body is rounded in shape, but for most of its length it is thin and flattened dorsoventrally. *Cerebratulus roseus* is anteriorly (the foregut regions) pink or light flesh coloured, posteriorly yellowish or with a faint reddish hue. In mature females the orange-brown colour of the eggs may give the body a darker appearance. The colourless lateral body margins are usually quite conspicuous.

Cerebratulus roseus has been found in limestone burrows and in sand from low water mark to about 30 m depth. Its geographic range extends from the English Channel (Fig. 18D) to the Mediterranean.

Genus *EUBORLASIA* Vaillant, 1890

Diagnosis

Heteronemertea with a single pair of short horizontal lateral cephalic slits which continue internally as deep intramuscular canals; proboscis containing two muscle layers (outer circular, inner longitudinal) and 1–2 muscle crosses; rhynchocoel wall circular muscles not interwoven with body wall musculature; dorsal fibrous core of cerebral ganglia forked only at rear into dorsal and ventral branches; nervous system with neither neurochords nor neurochord cells; foregut with circular splanchnic muscles; dermis divisible into distinct outer glandular and inner connective tissue zones; caudal cirrus absent; cephalic glands well developed; frontal organ present?; eyes absent; sexes separate; fat, bulky animals typically contracting into a coil.

Euborlasia elizabethae (McIntosh, 1873–74)
(Fig. 19)

Borlasia elizabethae McIntosh, 1873–74, 1875b; Riches, 1893; Sheldon, 1896
Euborlasia elisabethae Wijnhoff, 1912

Specific internal characters

Proboscis with one muscle cross; body wall in foregut regions with thin diagonal muscle layer between outer longitudinal and middle circular muscles.

Description

Euborlasia elizabethae reaches a length of up to about 30 cm and a width of 5–6 mm. When the animal is extended the body is rounded but on contraction it becomes distinctly flattened, especially in the intestinal region. The epidermis is generally very wrinkled, with both transverse and longitudinal corrugations evident over the body surface (Fig. 19A). The tail is rounded, the head tapered and bluntly pointed.

Most of the head is white or pale yellow in colour, speckled with olive-green or brown flecks of pigment. Towards the rear of the head and over the remaining dorsal body surface the general colour is a dark brown or reddish-brown speckled with white, yellow, light brown and olive-green and more or less regularly marked with transverse belts of pale pink or dirty white. The colours are brightest in the anterior parts of the body and gradually fade posteriorly. The ventral surface is similarly marked but with a paler hue.

Euborlasia elizabethae, which is not especially well described, has been found in the British Isles only once (Fig. 19B), on clayey mud under a stone in a rock pool near low water mark, but at other localities it has been dredged from depths of 4–50 m. The geographic range extends to the Mediterranean.

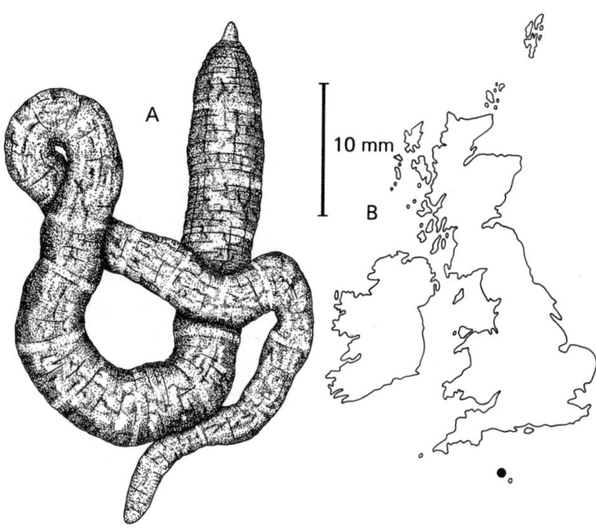

Fig. 19. *Euborlasia elizabethae*; A, general view of complete individual; B, recorded distribution from the British Isles. A redrawn from Joubin (1894) and McIntosh (1873–74).

Genus *LINEUS* Sowerby, 1806

Diagnosis

Heteronemertea with a single pair of horizontal lateral cephalic slits; proboscis containing two (outer circular, inner longitudinal) or three (outer longitudinal, middle circular, inner longitudinal) muscle layers and 0–2 muscle crosses; rhynchocoel wall circular musculature not interwoven with body wall muscle layers; dorsal fibrous core of cerebral ganglia branched only at rear into dorsal and ventral points; nervous system with neither neurochords nor neurochord cells; foregut with splanchnic musculature composed either of longitudinal fibres only or mixed longitudinal and circular layers; dermis variable, either with gland cells distributed among body wall outer longitudinal muscle fibres or separated from musculature by distinct connective tissue layer; caudal cirrus absent; cephalic glands usually well developed; most species with frontal organ; eyes present or absent; sexes separate.

Lineus acutifrons Southern, 1913
(Fig. 20A)

Lineus acutifrons Southern, 1913; Farran, 1915

Specific internal characters

None adequately described.

Description

Lineus acutifrons (Fig. 20A) is an Irish species which, since it has never been fully described, must remain of doubtful validity. The head, which lacks eyes but bears wide and deep lateral cephalic slits, is acutely pointed and distinctly marked off from the succeeding body regions by an obvious constriction. The ventral mouth appears just behind this constriction as a longitudinal slit. No intact individuals were collected and the largest body fragment measured 150–170 mm in length and 5–7 mm in width depending upon the degree of contraction. The worms are either pure white anteriorly, gradually becoming pink to bright red or brownish-purple posteriorly, or coral red in front and pink behind. There are no differences in colouration between the dorsal and ventral surfaces.

Lineus acutifrons lives in sand near low water mark. It has been found only in two regions of Blacksod Bay, Mayo, and in Ballynakill Harbour, Connemara (Fig. 22A).

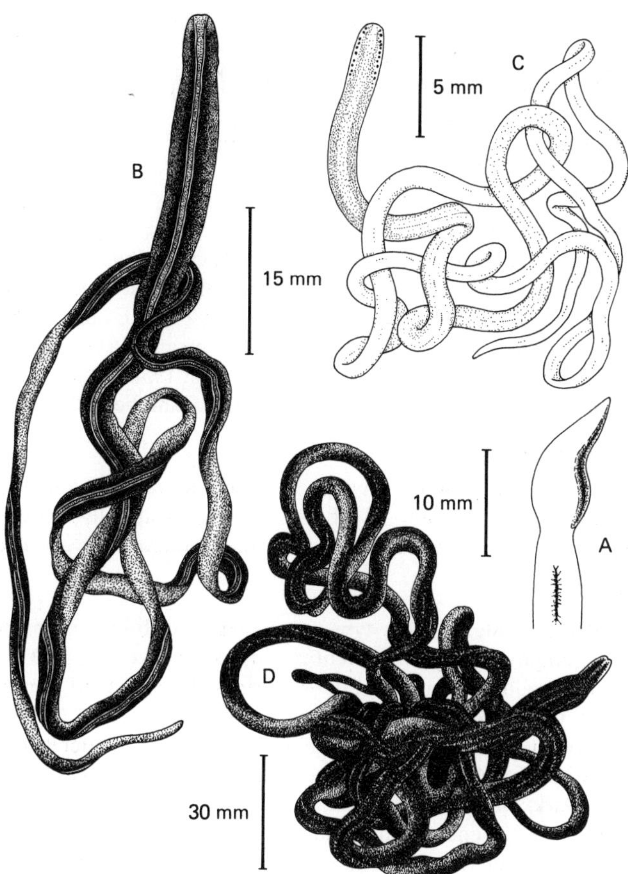

Fig. 20. *Lineus acutifrons*; A, ventral view of anterior regions, showing the large mouth and distinct lateral cephalic slits. *Lineus bilineatus*; B, general view of complete animal. *Lineus lacteus*; C, general view of complete animal. *Lineus longissimus*; D, general view of complete animal. A redrawn from Southern (1913).

Lineus bilineatus (Renier, 1804)
(Fig. 20B)

Cerebratulus bilineatus Renier, 1804; McIntosh, 1869
Gordius taenia Dalyell, 1853
Meckelia taenia Johnston, 1865; Lankester, 1866
Cerebratulus taenia McIntosh, 1868b, 1869
Cerebratulus oerstedtii Koehler, 1885
Lineus bilineata Todd, 1905
Lineus bilineatus McIntosh, 1873–74, 1875c; Riches, 1893; Gamble, 1896; Sheldon, 1896; Jameson, 1898; Allen, 1899; Allen & Todd, 1900; Beaumont, 1900a, b; M.B.A., 1904, 1931, 1957; Evans, 1909; King, 1911; Wijnhoff, 1912; Southern, 1913; Farran, 1915; Jones, 1939; Bassindale & Barrett, 1957; Barrett & Yonge, 1958; Bruce *et al.*, 1963; Crothers, 1966; Laverack & Blackler, 1974; Campbell, 1976

Specific internal characters

Proboscis with three muscle layers, but the outer longitudinal layer is reduced to isolated bundles of fibres arranged bilaterally; foregut with longitudinal splanchnic muscles; cephalic glands moderately well developed.

Description

Lineus bilineatus (Fig. 20B) is typically a rich reddish-brown or chocolate colour, sometimes paler on the ventral surface, marked with two slender white or pale yellow longitudinal dorsal stripes which extend the full body length, but the background colouration is quite variable. Younger and smaller individuals are often much paler in colour than older animals and may appear almost white. In larger specimens the colour ranges from shades of pink to dark purple. *Lineus bilineatus* attains a length of 45–50 cm or more. The body is slender and gradually tapers towards the posterior end. The head is bluntly rounded and often rather flattened and spatulate in shape. It does not possess eyes.

McIntosh (1873–74) records that though easily maintained in captivity, the species is sluggish in its habits and may remain quiescent for several weeks under a shell or ensheathed in mucus.

A lower middle shore to sublittoral species, *Lineus bilineatus* is found in the British Isles (Fig. 22B) in shelly gravel, muddy sand or clean fine sand, amongst coralline algae in rock pools, under *Laminaria* holdfasts, beneath stones and boulders, or between mussel or oyster shells. It has a geographic distribution extending from Iceland and the Faroe Islands to Scandinavia, France, the Mediterranean and Madeira. It has also been reported from the west coast of North America (Alaska to California), South Africa and Japan. The Japanese record (Iwata, 1954) almost certainly refers to another species as it possesses both a different colour pattern and six eyes.

Lineus lacteus (Rathke, 1843)
(Fig. 20C)

Ramphogordius lacteus Rathke, 1843
? *Gordius minor albus* Dalyell, 1853
? *Lineus albus* Johnston, 1865
Borlasia lactea Parfitt, 1867; McIntosh, 1868a, 1869
Lineus lactea McIntosh, 1867b, 1869
Lineus lacteus Montagu, 1808; McIntosh, 1873–74; Riches, 1893; Sheldon, 1896; Beaumont, 1900a, b; M.B.A., 1904, 1931, 1957; Southern, 1908a; Wijnhoff, 1912

Specific internal characters
Dermis without connective tissue layer; proboscis extraordinarily slender, with two muscle layers.

Description
Lineus lacteus (Fig. 20C) attains a length of 30–60 cm but is only 1–2 mm wide. The elongate head is marginally broader than the succeeding trunk regions and bears from 6–15 eyes in a dorsolateral row on each side. The mouth is located some distance behind the cerebral ganglia in contrast to the usual immediate post-cerebral position in other lineids. Behind the cerebral ganglia the pale pink to reddish colouration of the head gradually fades to a uniform white or pale creamish-yellow; the tip of the snout and the tail are translucent.

Lineus lacteus is apparently much more common on Mediterranean than on Atlantic coasts and in the British Isles has been found only in south-western Ireland and on the south and west coasts of England and Wales respectively (Fig. 22A). The *Lineus albus* recorded from Scottish and Cornish coasts by Johnston (1865) may be an *Amphiporus* species according to Bürger (1895), although McIntosh (1873–74) observes that 'Dr Johnston's preparation of *Lineus albus* . . . resembles the present species very closely.'

The geographic distribution extends from the British and French Atlantic coasts to the Mediterranean and Black Seas. The species has also been reported from Sweden, a record doubted by Wijnhoff (1912). *Lineus lacteus* has been found sublittorally to intertidally, on gravel, sand or beneath stones.

Lineus lacteus is apparently hardy in captivity, as McIntosh (1873–74) records having kept examples alive for more than five years. Hermaphroditic individuals, with the posterior body full of ripe-looking ova and the anterior containing mature sperm, are listed by M.B.A. (1931, 1957), although hermaphroditism is not a feature of the genus *Lineus*.

Lineus longissimus (Gunnerus, 1770)
(Fig. 20D)

Sea Long Worm Borlase, 1758
Ascaris longissima Gunnerus, 1770
Gordius marinus Montagu, 1804; Pennant, 1812
Nemertes borlasii Griffith, 1834; Thompson, 1838; Kingsley, 1859
Borlasia longissimus Templeton, 1836
Nemertes gracilis Goodsir, 1845
Lineus gracilis Johnston, 1846, 1865; Lankester, 1866
Gordius maximus Dalyell, 1853
'low' Ascarid or Planarian worm Anonymous, 1854
Borlasia nigra Byerley, 1854a, b
Nemertes borlassii Beattie, 1858; Logan, 1860
Borlasia striata Johnston, 1865
Lineus lineatus Johnston, 1865; Lankester, 1866; Parfitt, 1867
Lineus murenoides Johnston, 1865
Lineus fasciatus Johnston, 1865
Lineus marinus Montagu, 1808; McIntosh, 1873–74, 1875c, 1876; Haddon, 1886a, b; Gibson, 1886; Scott, 1893; Duerden, 1894; Herdman, 1896, 1903; Sheldon, 1896; Allen, 1899; Gemmill, 1901; Newbigin, 1901; Chumley, 1918
Lineus (Gordius) marinus Evans, 1909
Lineus longissimus Sowerby, 1806; Neill, 1807; Turton, 1807; Jameson, 1811; Davies, 1815, 1816, 1817; Thompson, 1843, 1856; Johnston, 1846, 1865; Williams, 1852; Lankester, 1866; Parfitt, 1867; McIntosh, 1868a, b, 1869, 1870; Koehler, 1885; Riches, 1893; Herdman, 1894, 1900; Beaumont, 1895a, b, 1900a, b; Vanstone & Beaumont, 1894, 1895; Gamble, 1896; Jameson, 1898; Allen & Todd, 1900; M.B.A., 1904, 1931, 1957; King, 1911; Stephenson, 1911; Wijnhoff, 1912; Southern, 1913; Walton, 1913; Farran, 1915; Hunt, 1925; Renouf, 1931; Moore, 1937; Stopford, 1951; Eales, 1952; Bassindale & Barrett, 1957; Barrett & Yonge, 1958; Clark & Cowey, 1958; Bruce *et al.*, 1963; Crothers, 1966; Laverack & Blackler, 1974; Campbell, 1976; Cantell, 1976; Boyden *et al.*, 1977; Knight-Jones & Nelson-Smith, 1977

Specific internal characters
Dermis with distinct connective tissue layer; proboscis with two muscle layers and two muscle crosses.

Description
Lineus longissimus is the longest nemertean species known. Individuals of 5–10 m length are not uncommon and McIntosh (1873–74) records a specimen from St Andrews of which 'Thirty yards were measured without rupture, and yet the mass was not half uncoiled.' The species possesses a flaccid body which, when disturbed, contracts and extends in a series of

irregular muscular waves. When handled the animals produce copious amounts of a rather viscid mucus which has a faintly pungent odour. The colour ranges from a dark olive-brown or rich chocolate brown in younger individuals to a blackish-brown or black in larger animals. Often a flickering purplish iridescence is evident over the body surface which is due to the activity of the epidermal cilia. The body may appear streaked with pale longitudinal lines (Fig. 20D), especially on the anterior dorsal surface, and the lateral margins of individuals containing mature gonads are often a pale greenish-brown hue. Ventrally the colour may be the same as or slightly paler than that of the dorsal surfaces. The cerebral ganglia show pink to red through the epidermis. The tip of the rectangular head is pale or whitish in colour and slightly bilobed. There are 10–20 deep-set reddish-brown or black eyes arranged in a row on each side of the snout.

Probably the most frequently recorded of the British nemertean species (Fig. 22C), *Lineus longissimus* is typically found on the lower shore coiled into gently writhing knots beneath boulders on muddy sand, but also occurs in rock-pools, entangled among laminarian holdfasts, in rock fissures and clefts, or in deeper sublittoral locations on muddy, sandy, stony or shelly bottoms. The geographic range of the species extends from Iceland eastwards to the Atlantic, North Sea and Baltic coasts of Europe, but it has not been found in the Mediterranean.

Lineus ruber (Müller, 1774)

Lineus sanguineus (Rathke, 1799)

Lineus viridis (Müller, 1774)

This group of species has given rise to much taxonomic confusion and each of the three specific names, together with a fourth (*gesserensis*), has at some time been used as a name for the others. Accordingly no attempt is made to separate the following synonyms into species groups, although in general the specific names *bioculata*, *gesserensis*, *obscurus*, *olivacea* and *purpurea* have been more widely applied to *Lineus ruber* or *Lineus viridis*, whereas the name *octoculata* is more typically associated with *Lineus sanguineus*.

Fasciola rubra Müller, 1774
Fasciola viridis Müller, 1774
Planaria sanguinea Rathke, 1799
? *Gordius oculatus* Montagu, 1802, 1808
? *Lineus oculatus* Montagu, 1808
? *Planaria rufa* Montagu, 1808
? *Lineus rufus* Montagu, 1808
? *Planaria fusca* Montagu, 1808; Thompson, 1843
? *Planaria unicolor* Johnston, 1827–28
Planaria bioculata Johnston, 1828–29
Planaria octoculata Johnston, 1828–29
Nemertes Borlasia octoculata Johnston, 1837
Nemertes Borlasia olivacea Johnston, 1837
Nemertes Borlasia purpurea Johnston, 1837
? *Borlasia? unicolor* Johnston, 1846
Borlasia octoculata Thompson, 1846, 1856; Johnston, 1846, 1865; Lankester, 1866; McIntosh, 1867b, 1868a, 1869, 1870; Gibson, 1886
Borlasia olivacea Thompson, 1846, 1856; Johnston, 1846, 1865; Lankester, 1866; McIntosh, 1867b, 1868a, 1869
Borlasia purpurea Thompson, 1846, 1856; Johnston, 1846, 1865; Parfitt, 1867
Gordius gesserensis Dalyell, 1853
Gordius minor viridis Dalyell, 1853
Nemertes octoculata Byerley, 1854a, b
Borlasia gesserensis Johnston, 1865
Lineus viridis Johnston, 1865; Cantell, 1975
'*Borlasia*' *purpurea* McIntosh, 1867a
Lineus gesserensis McIntosh, 1873–74, 1875b, c, 1876, 1927; Koehler, 1885; Herdman, 1894; Sheldon, 1896; Jameson, 1898; Beaumont, 1900a, b; Gemmill, 1901; Punnett, 1901b; M.B.A., 1904; Evans, 1909; Colgan, 1916; Elmhirst, 1922; Moore, 1937; Bassindale, 1941, 1943; Corlett, 1947; Purchon, 1948, 1956; Eales, 1952; Judges & Southward, 1953; Southward, 1953; Newell, 1954; Bassindale & Barrett, 1957; Barrett & Yonge, 1958; Clark & Cowey, 1958; Milne & Dunnet, 1972; Williams,

1972; Eason, 1973; Laverack & Blackler, 1974

Lineus obscurus Riches, 1893; Vanstone & Beaumont, 1894, 1895; Herdman, 1894, 1900; Beaumont, 1895a, b; Gamble, 1896

Lineus ruber Southern, 1908a, 1913; Stephenson, 1911; Wijnhoff, 1912; Farran, 1915; M.B.A., 1931, 1957; Evans, 1949; Eales, 1952; Bassindale & Barrett, 1957; Barrett & Yonge, 1958; Jennings, 1960, 1962, 1969; Markowski, 1962; Bruce *et al.*, 1963; Crothers, 1966; Gibson & Jennings, 1967; Gibson, 1968a; Green, 1968; Jennings & Gibson, 1969; Ling, 1969, 1970, 1971; Milne & Dunnet, 1972; Eason, 1973; Ling & Willmer, 1973; Laverack & Blackler, 1974; Slinger & Gibson, 1974, 1975; Cantell, 1975; Slinger, 1975; Campbell, 1976; Boyden *et al.*, 1977

Lineus sanguineus McIntosh, 1873–74, 1875b, c; Gemmill, 1901; Gibson & Jennings, 1967; Gibson, 1968a; Jennings & Gibson, 1969; Eason, 1973

Lineus ruber
(Fig. 21A, B)

Specific internal characters

Dermis without distinct connective tissue layer; cephalic glands well developed; proboscis with two muscle layers and two muscle crosses; body wall musculature without diagonal layer.

Description

Lineus ruber (Fig. 21A) is mostly less than 7–8 cm in length, although larger individuals may occasionally be encountered; McIntosh (1873–74), for example, records '*Lineus gesserensis*' as being from 'Four to nine inches in length'. Specimens longer than 10–12 cm must be regarded as very uncommon. The head is bluntly rounded at the front, with two to eight small brown to black eyes forming an irregular row on each dorsolateral margin. In juveniles there is often only a single pair of eye-spots. Behind the lateral cephalic slits the body is of a more or less uniform width and is somewhat dorsoventrally flattened. In colour *Lineus ruber* is typically a light to dark reddish-brown, with a paler ventral surface. The cephalic margins and sides of the body often appear a translucent white. Variations in the colouration include violet, greenish-red and yellowish-brown forms, and specimens in which the pigmentation is darkest in the anterior regions. The cerebral ganglia usually show pink or red through the dorsal surface. In sexually mature animals the gonads appear as whitish spots along the lateral intestinal margins.

When mechanically irritated *Lineus ruber* contracts without coiling into a spiral; this behaviour enables it to be distinguished from *Lineus sanguineus* which contracts in a tightly coiled fashion.

Lineus ruber reproduces from January to May, laying gelatinous egg-strings which may commonly be found adhering to the underside of rocks and boulders. It is an exceptionally hardy species in captivity and individuals have been kept without aeration or food for up to twelve months. A very common British nemertean (Fig. 22D), *Lineus ruber* is typically found intertidally on muddy sand beneath stones and boulders from the *Fucus spiralis* level downwards. Often several individuals can be found under a single rock. It is also found in mussel beds, crawling between barnacles, on rock-pool algae (especially *Cladophora* and *Ceramium*), among the holdfasts of the larger brown seaweeds, in estuarine muds and sublittorally on a variety of bottom types. Not uncommonly occurring in brackish waters, the lower limit of salinity tolerance of the species appears to be about 8‰ (Remane, 1958).

Lineus ruber is one of the most commonly recorded nemerteans and has a circumpolar distribution in the northern hemisphere, being reported from Atlantic, Mediterranean and North Sea European coasts, both Pacific and Atlantic coasts of North America, Madeira, Greenland, Iceland, the Faroes and the coast of Siberia. It is also recorded from South Africa.

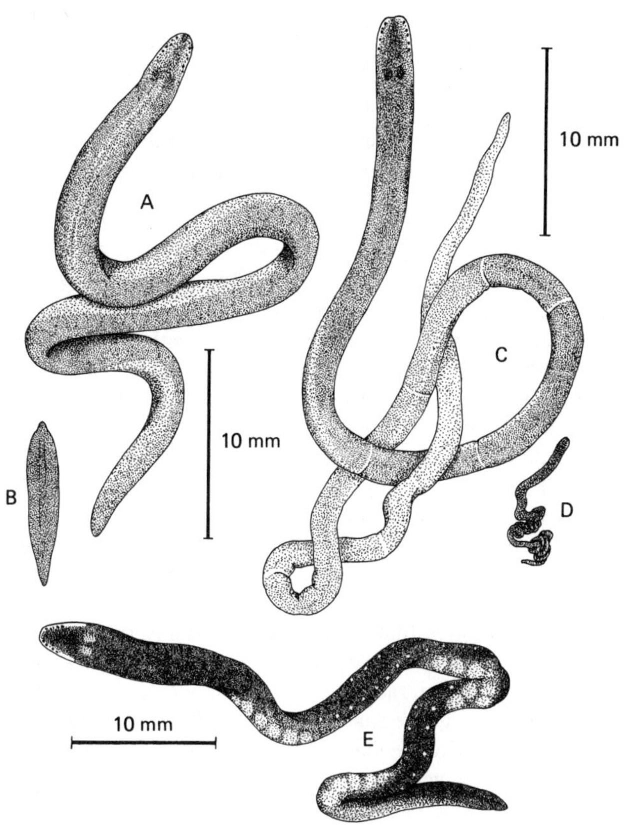

Fig. 21. *Lineus ruber*; A, general view of whole animal; B, dorsal view of fully contracted individual. *Lineus sanguineus*; C, general view of complete individual; D, a fully contracted specimen, showing the characteristic spiral coiling of the body. *Lineus viridis*; E, general view of whole animal.

Lineus sanguineus
(Fig. 21C, D)

Specific internal characters

Dermis without distinct connective tissue layer; proboscis with two muscle layers and two muscle crosses; body wall musculature with slender diagonal layer.

Description

Somewhat similar to *Lineus ruber* in appearance, *Lineus sanguineus* is up to about 10–20 cm in length but possesses a much more slender body (Fig. 21C). On the head the four to six eyes on each side are rather further back than in *Lineus ruber* and more regularly arranged. The colour varies from a bright reddish-brown to a dull mid-brown, frequently the posterior and ventral regions appearing paler. The species is readily distinguished from *Lineus ruber* by its behaviour when disturbed, contracting in a tight coil (Fig. 21D) (p. 93).

Lineus sanguineus is found intertidally beneath rocks and stones embedded in muddy sediments from the mid-shore level downwards, often in muds blackened by decaying organic matter. According to McIntosh (1873–74) *Lineus sanguineus* reproduces sexually during the autumn, whilst Gontcharoff (1951) notes that it multiplies asexually throughout the year. The species possesses well developed powers of regeneration.

The known geographic range of *Lineus sanguineus* extends from the British Isles (Fig. 23A) to the coasts of Sweden, Belgium and France, though a wider distribution may be masked by the problems of synonymity. For example, *Lineus sanguineus* on the Maine coast of the United States (Verrill, 1892) is regarded by Coe (1943) as *Lineus ruber*.

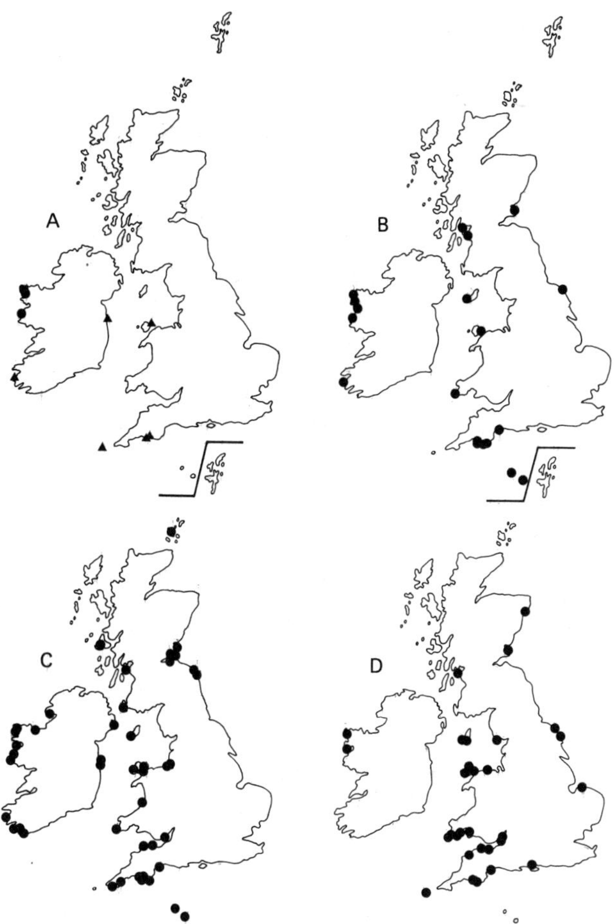

Fig. 22. A, recorded distribution of *Lineus acutifrons* (●) and *Lineus lacteus* (▲) from the British Isles; B, recorded distribution of *Lineus bilineatus* from the British Isles; C, recorded distribution of *Lineus longissimus* from the British Isles; D, recorded distribution of *Lineus ruber* from the British Isles.

Lineus viridis
(Fig. 21E)

Specific internal characters

Dermis without distinct connective tissue layer; cephalic glands well developed; proboscis with two muscle layers and two muscle crosses; body wall musculature without diagonal layer.

Description

In shape, size and eye arrangement *Lineus viridis* (Fig. 21E) is generally very similar to *Lineus ruber*. It differs, however, in being a dark olive-green to greenish-black colour, although pale green individuals are sometimes found and in these the reddish cerebral ganglia are particularly evident through the dorsal body wall. The ventral surface may be marginally lighter in hue than the dorsal. Mature adults are sometimes marked with pale-coloured slender annulations arranged at more or less regular intervals. In sexually ripe specimens the gonads and gonopores appear as obvious whitish spots arranged in a dorsolateral row on either side of the body. Eggs are laid during April.

Lineus viridis, most commonly under the name *Lineus gesserensis*, was for long regarded merely as a colour variety of *Lineus ruber* and not as a separate species. Gontcharoff (1951, 1959, 1960), however, demonstrated that the two forms differ strikingly in their larval development and she accordingly regarded them as distinct species. In *Lineus ruber* few (10–15) larvae of very variable size emerge from the egg-string after hatching and commence feeding immediately. They also exhibit a weakly positive, or, more usually, indifferent response to light which rapidly develops into a strong negative phototaxis. In contrast 400–500 small larvae emerge from the egg-strings of *Lineus viridis* and possess an initial period of strongly positive phototropism, lasting two to three weeks or more, during which they do not require food. The juveniles of both species are transparent and eyeless; in *Lineus ruber* the body pigmentation begins to appear after two to three months, but in the slower-growing *Lineus viridis* the colour takes longer to develop.

The geographic distribution of *Lineus viridis* closely parallels that of *Lineus ruber* except that it is not known outside the northern hemisphere. The species does not appear to be as common as *Lineus ruber* (Fig. 23B), although it may be locally abundant. It is most often found on muddy sediments beneath boulders or stones from the mid-shore (*Fucus vesiculosus*, *Ascophyllum nodosum*) level downwards, but may occur higher and on somewhat coarser substrata. *Lineus viridis* is also reported from estuaries.

Some authorities credit Fabricius (1780) as the naming author for this species; Fabricius, however, lists Müller's (1774) *Fasciola viridis* as a synonym for his *Planaria viridis* and for this reason *Lineus viridis* is here considered to have been named originally by Müller.

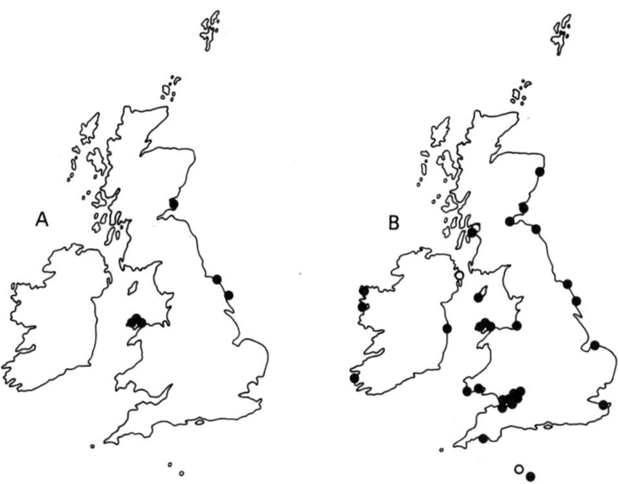

Fig. 23. A, recorded distribution of *Lineus sanguineus* from the British Isles; B, recorded distribution of *Lineus viridis* from the British Isles; ● = certain or very probable records; ○ = records which may be of either *Lineus ruber* or *Lineus viridis*.

Genus *MICRELLA* Punnett, 1901

Diagnosis

Heteronemertea with a single pair of horizontal lateral cephalic slits; proboscis with two muscle layers (outer circular, inner longitudinal) and two muscle crosses; rhynchocoel wall circular musculature not interwoven with body wall muscle layers; rhynchocoel in foregut region with shallow lateral diverticula, lacking muscular walls, which are in close communication with branches of foregut vascular plexus; dorsal fibrous core of cerebral ganglia branched only at rear into upper and lower points; nervous system with neither neurochords nor neurochord cells; dermal gland cells situated amongst body wall outer longitudinal muscle fibres; body wall musculature without diagonal layer; caudal cirrus present; dorsoventral muscles present but poorly developed; cephalic glands weakly developed; no eyes or frontal organ, but with a pair of lateral sensory organs situated just behind excretory pores; sexes probably separate.

Micrella rufa Punnett, 1901
(Fig. 24)

Micrella rufa Punnett, 1901a; M.B.A., 1904, 1931, 1957

Specific internal characters

The genus *Micrella* is monospecific; the specific internal characters are thus as given in the generic diagnosis.

Description

Known from only two specimens, one incomplete, dug from mud at low tide level near the mouth of the River Yealm, Plymouth (Fig. 24B), *Micrella rufa* (Fig. 24A) has an elongate, slender body up to 18 cm or more long and 2–3 mm wide. Posteriorly it is flattened dorsoventrally and terminates in a delicate caudal cirrus, above the base of which the anus discharges. The head is pointed and bears shallow cephalic slits. The body is a uniform bright vermilion red colour, shading into a yellowish tinge near the tip of the head. The cerebral ganglia, visible through the body wall, are bright red, and in the intestinal regions the gut and lateral diverticula appear brownish.

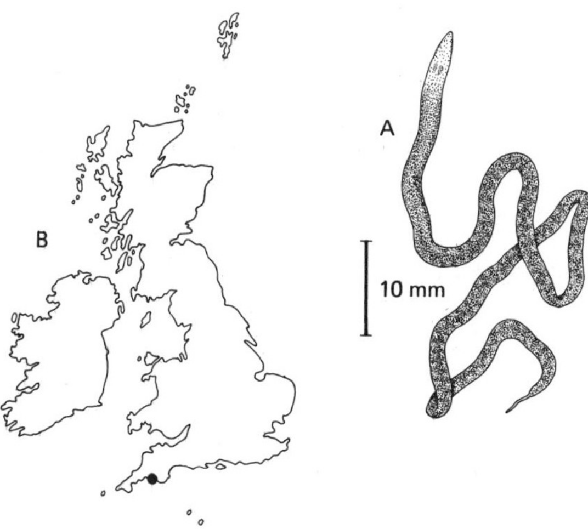

Fig. 24. *Micrella rufa*; A, general view of complete individual; B, recorded distribution from the British Isles. A based upon description given by Punnett (1901a).

Genus *MICRURA* Ehrenberg, 1831

Diagnosis

Heteronemertea with a single pair of horizontal lateral cephalic slits, posteriorly enlarged to form wide bays; ciliated cerebral canals emerge from ventral wall of cephalic bays; proboscis containing two (outer circular, inner longitudinal) or three (outer longitudinal, middle circular, inner longitudinal) muscle layers and 0–2 muscle crosses; rhynchocoel wall circular muscles not interwoven with body wall musculature; dorsal fibrous core of cerebral ganglia branched into upper and lower forks only at rear; nervous system with neither neurochords nor neurochord cells; ganglionic cell layer of brain lobes usually not separated from body wall musculature by distinct neurilemma; foregut with or without splanchnic muscles, if present variably composed of circular and/or longitudinal fibres; dermis variable, usually with distinct connective tissue zone separating glandular region from body wall musculature; caudal cirrus present; cephalic glands normally well developed, occasionally weakly formed or absent; frontal organ usually present; eyes present or absent; sexes separate.

Micrura aurantiaca (Grube, 1855)
(Fig. 25A)

Meckelia aurantiaca Grube, 1855

Micrura aurantiaca McIntosh, 1873–74; Riches, 1893; Sheldon, 1896; M.B.A., 1904, 1931, 1957; Wijnhoff, 1912; Eales, 1952; Campbell, 1976

Specific internal characters

Dermis without distinct connective tissue layer; body wall inner longitudinal musculature reduced in development.

Description

Rarely more than about 6–8 cm long and 1.5–2.0 mm wide, *Micrura aurantiaca* (Fig. 25A) possesses a body which is dorsally rounded and ventrally flattened. The head is slightly tapered and blunt and lacks eyes. At the posterior end the translucent caudal cirrus is small and often indistinct. In colour *Micrura aurantiaca* is typically a bright brick red dorsally, white or pink ventrally. On the head a white patch separates an anterior dorsal cephalic spot of reddish, violet or brownish pigmentation which is variable both in size and shape; exceptionally it may be completely missing. The extreme tip of the head is also white.

More often obtained sublittorally from coralline grounds, *Micrura aurantiaca* may also be found intertidally beneath stones in rock pools. The geographic range of the species extends from the British Isles (Fig. 26A) to the Mediterranean.

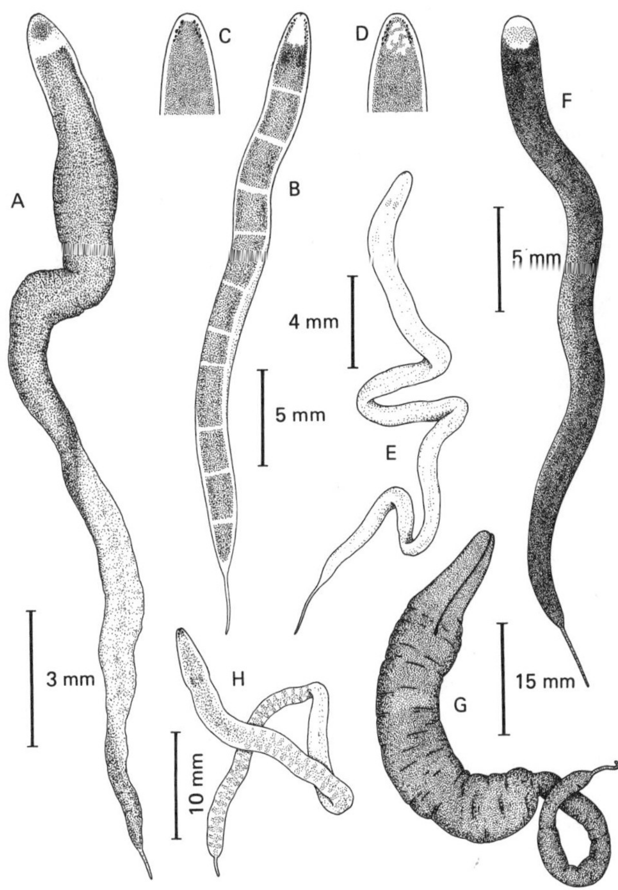

Fig. 25. *Micrura aurantiaca*; A, general view of complete individual. *Micrura fasciolata*; B, dorsal view of whole animal; C, D, dorsal views of the heads of two specimens to show variations in the amount of white cephalic pigmentation. *Micrura lactea*; E, general view of whole animal. *Micrura purpurea*; F, dorsal view of whole animal. *Micrura rockalliensis*; G, lateral view of preserved individual. *Micrura scotica*; H, general view of complete specimen. A, E redrawn from Bürger (1895), C, D redrawn from Cantell (1975), G redrawn from Dollfus (1924), H redrawn from Stephenson (1911).

Micrura fasciolata Ehrenberg, 1831
(Fig. 25B, C, D)

Planaria lineata Montagu, 1808
Gordius fasciatus spinifer Dalyell, 1853
Stylus fasciatus Johnston, 1865
Micrura fasciolata Ehrenberg, 1831; McIntosh, 1869, 1873–74, 1875c; Riches, 1893; Beaumont, 1895a, b, 1900a, b; Gamble, 1896; Sheldon, 1896; Herdman, 1896, 1900; Jameson, 1898; Allen, 1899; Allen & Todd, 1900; M.B.A., 1904, 1931; Southern, 1908a, 1913; King, 1911; Wijnhoff, 1912; Farran, 1915; Moore, 1937; Jones, 1939; Bruce *et al.*, 1963; Laverack & Blackler, 1974
Micrura fasiolata Bassindale & Barrett, 1957; M.B.A., 1957; Crothers, 1966

Specific internal characters

Dermis with distinct connective tissue layer; proboscis with two muscle layers and two muscle crosses; cephalic glands well developed.

Description

A strikingly coloured species, *Micrura fasciolata* (Fig. 25B) attains a length of 10–15 cm or more but is only 1–4 mm wide. The bluntly rounded head, which tapers anteriorly, bears distinct cephalic slits and 3–12 small reddish-brown to blackish eyes arranged in a row near each cephalic margin. Eye number varies with size and age; in juvenile worms with a length of about 1–2 mm there is usually only a single pair of distinct eye-spots situated near the tip of the head. The body is somewhat flattened ventrally and gently tapers posteriorly to end in a pale caudal cirrus which may be quite long.

The colour pattern is characteristic for the species. Dorsally *Micrura fasciolata* is usually a rich reddish-brown hue marked with regularly or irregularly distributed white transverse bars, but the background colour may be brown, yellowish-brown, greenish-brown or red-violet. Sometimes the white bars are indistinct or lozenge-shaped. The tip of the snout and the cephalic slits are white. A dorsal but very variable patch of pigmentation is present on the head, which may be entirely coloured (apart from the white tip) or be predominantly white (Fig. 25C, D). The cephalic pigment may also be slightly darker than that of the remaining dorsal surface. Ventrally the species is a much paler colour, although still tending to be reddish-brown and marked with faint lines or furrows which are continuous with the white dorsal bands. Younger individuals generally possess a paler colouration, and often the posterior portion of the body is uniformly tinted without transverse markings.

Micrura fasciolata is often locally abundant in sublittoral situations, dredged at depths to 80 m or more, but may occasionally be found intertidally near low water level. The species has been found in a wide range of sediments, including sand, gravel, shell fragments and mud but is also commonly associated with tubicolous polychaete tubes, especially

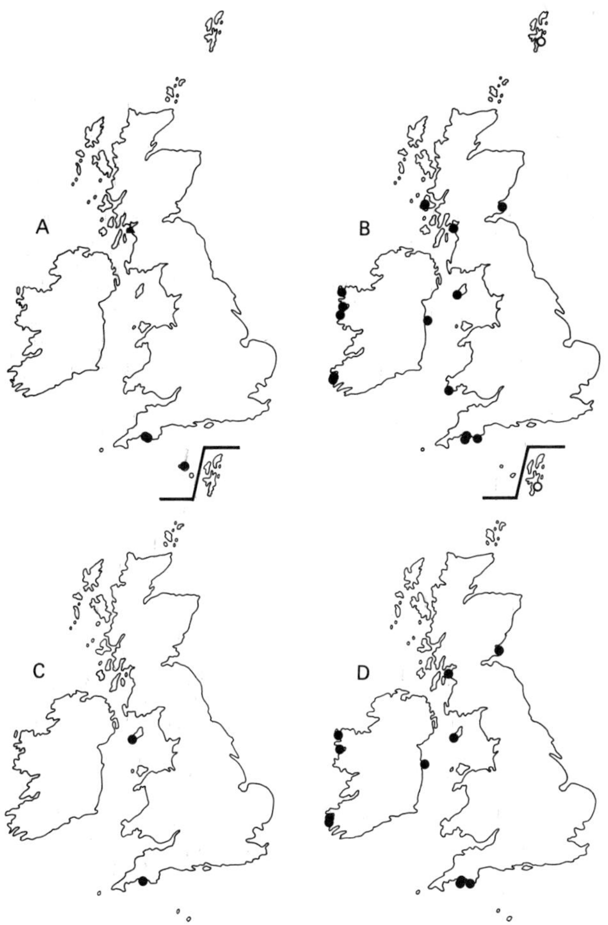

Fig. 26. A, recorded distribution of *Micrura aurantiaca* (●) and *Micrura scotica* (▲) from the British Isles; B, recorded distribution of *Micrura fasciolata* from the British Isles; ● = certain records, ○ = specific identity unconfirmed; C, recorded distribution of *Micrura lactea* from the British Isles; D, recorded distribution of *Micrura purpurea* from the British Isles; ● = certain records, ○ = specific identity unconfirmed.

Pomatoceros, laminarian holdfasts and *Zostera* beds or may be found in rocky fissures or beneath boulders and stones. Sexually mature individuals are generally obtained during the September to December period, rather later in more northerly latitudes.

The geographic range of *Micrura fasciolata* extends from Scandinavia to the British Isles (Fig. 26B) and the Mediterranean.

Micrura lactea (Hubrecht, 1879)
(Fig. 25E)

Cerebratulus lacteus Hubrecht, 1879
Micrura candida Riches, 1893; Beaumont, 1895a, b; Herdman, 1896, 1900; Sheldon, 1896; Wijnhoff, 1912; M.B.A., 1931, 1957; Moore, 1937; Bruce *et al.*, 1963
Micrura lactea M.B.A., 1904

Specific internal characters

Dermis without distinct connective tissue layer; cephalic glands very well developed.

Description

Despite being most commonly recorded as *Micrura candida*, the name *Micrura lactea* has priority for this species. With a maximum length of about 8 cm, and width of 1.0–1.5 mm, *Micrura lactea* (Fig. 25E) possesses an attenuated and gradually tapering body which posteriorly ends in a delicate and very slender caudal cirrus. The head is bluntly rounded, dorsoventrally flattened and lacks eyes, and is not obviously delimited from the trunk.

Typically an overall milk-white colour, the gut contents may give the intestinal regions a pale yellowish or brownish flesh-pink hue, and occasional examples may be more uniformly tinged a pale rose-pink. The cerebral ganglia may show as a pair of darker pink lobes at the rear of the head. Under a microscope the epidermis may appear to be covered in opaque white flakes due to the large size of many of the gland cells.

Micrura lactea, with a geographic range extending from the British Isles (Fig. 26C) to the Mediterranean, has only rarely been found in British waters. Examples have so far been obtained only by dredging at depths to about 30 m.

Micrura purpurea (Dalyell, 1853)
(Fig. 25F)

Gordius purpureus spinifer Dalyell, 1853
Stylus purpureus Johnston, 1865; McIntosh, 1868a
Micrura (Stylus) purpurea McIntosh, 1869
Micrura purpurea McIntosh, 1873–74, 1875c; Riches, 1893; Beaumont, 1895a, b, 1900a, b; Gamble, 1896; Herdman, 1896; Sheldon, 1896; Jameson, 1898; Allen, 1899; Gemmill, 1901; M.B.A., 1904, 1931, 1957; Southern, 1908a, 1913; Wijnhoff, 1912; Farran, 1915; Moore, 1937; Jones, 1939; Bruce *et al.*, 1963; Laverack & Blackler, 1974

Specific internal characters

Dermis with distinct connective tissue layer; proboscis with two muscle layers and two muscle crosses; thick dorsoventral muscles developed between intestinal diverticula.

Description

Up to 20 cm or more long and 2–3 mm wide, *Micrura purpurea* (Fig. 25F) possesses a slightly flattened body which gradually tapers posteriorly to end in a delicate and slender caudal cirrus. The head is bluntly rounded, lacks eyes and is not distinctly marked off from the remainder of the body.

The colour is a dark rich purplish-brown on the dorsal surface, in life often appearing iridescent when light is reflected by the beating epidermal cilia. The head is generally darker, with the cerebral ganglia visible through the body wall as a pair of reddish lobes, but is anteriorly marked by a transverse band of brilliant yellow or yellowish-white which is characteristic of the species. On occasion this band may be divided into two yellow spots, rarely it is completely absent. In front of the band the snout is whitish or translucent, behind it there are usually a few white granules. The ventral body surface is a similar colour to the dorsal, although it is frequently somewhat paler in hue. The caudal cirrus too is pigmented, usually a pale brown colour.

Micrura purpurea is often found associated with the related species, *Micrura fasciolata*, and occurs in an equally wide range of habitat types. It is most commonly obtained by dredging muddy, shelly, sandy or gravelly sediments at depths of 10–40 m, but may dwell intertidally near low water level in rocky clefts, among laminarian holdfasts, beneath stones and boulders, or between the shells of bivalve molluscs. Sexually mature individuals are found during the April to June period. In captivity *Micrura purpurea* demonstrates voracious feeding habits and readily devours worms of various types; it is sometimes cannibalistic.

The geographic distribution of *Micrura purpurea* parallels that of *Micrura fasciolata* (Fig. 26D).

Micrura rockalliensis Dollfus, 1924
(Fig. 25G)

Micrura rockalliensis Dollfus, 1924

Specific internal characters

Not known; the internal anatomy of this species has never been described.

Description

An inadequately described species of dubious validity, known from only a single individual dredged from 175 m depth at 58°N latitude and 13°55′W longitude (near Rockall) in the Atlantic Ocean. The specimen was 5.5 mm long, 0.9 mm wide, with a rounded body posteriorly terminating abruptly in a distinct caudal cirrus (Fig. 25G). The sharply tapering head, bluntly rounded at its anterior tip and without eyes, was somewhat flattened dorsoventrally. The colour in life is unknown; after preservation in alcohol the body assumed a uniform blackish-brown hue.

Micrura scotica Stephenson, 1911
(Fig. 25H)

Micrura scotica Stephenson, 1911; King, 1911

Specific internal characters

Not known; the species is inadequately described.

Description

The only known example of *Micrura scotica* (Fig. 25H) is 6.5 cm long and 2.5 mm wide, with a posteriorly tapering body which is oval in cross-section. The bluntly rounded head has an elongate triangular shape, with distinct cephalic slits whose posterior regions are red in colour. The eyes are arranged in two lateral rows, with about 5–8 in each; the anterior eye of each row is larger and more distinct than the rest. At the posterior end of the body the whitish caudal cirrus, 0.8 mm long, is quite evident in life.

The dorsal colour is a uniform light brown anteriorly, with a purplish tinge, but in the intestinal regions only the gut and its lateral diverticula appear similarly pigmented and the remaining parts of the body have a paler hue. The margins of the head, mouth and body are white, the ventral surface whitish. A small reddish patch appears near the tip of the head, between the anterior pair of eyes, and the cerebral ganglia too show as a reddish region on the dorsal body surface.

Micrura scotica was dredged from about 30 m depth in the Firth of Clyde (Fig. 26A).

Family POLIOPSIIDAE

Genus *POLIOPSIS* Joubin, 1890

Diagnosis

Heteronemertea without lateral head slits but with dorsal and ventral median longitudinal cephalic furrows; proboscis slender, with two (outer longitudinal, inner circular) muscle layers; dorsal fibre core of cerebral ganglia forked only at rear into upper and lower branches; dermis with distinct glandular and connective tissue layers; caudal cirrus absent; cerebral sensory organs large and well developed, attached to rear of dorsal cerebral ganglionic lobes; frontal organ present; lateral nerve cords extend posteriorly inside body wall circular muscle layer; intestine with short ventral caecum extending below pylorus-like posterior region of foregut; head small, conical, separated from body by distinct but dorsally incomplete serrate annular constriction; eyes numerous; sexes separate.

Poliopsis lacazei Joubin, 1890
(Fig. 27)

Poliopsis lacazei Joubin, 1890; Wijnhoff, 1912; M.B.A., 1931, 1957

Specific internal characters

The genus *Poliopsis* is monospecific; the specific internal characters are thus as given in the generic diagnosis.

Description

Poliopsis lacazei (Fig. 27A) attains a length of 40–50 cm and width of 5–8 mm, although most individuals have been considerably smaller than this. The distinct head bears up to about 80 small black eyes, arranged in loose dorsolateral groups on either side of the median dorsal furrow (Fig. 27B). The body is bulky, circular in section and when contracted possesses a wrinkled epidermal surface. Bright pink to greyish-red in colour anteriorly, the alimentary tract imparts a yellowish hue to the intestinal regions where the body wall is more or less transparent. The tip of the head is translucent and virtually colourless.

The only known member of this genus, *Poliopsis lacazei* has been found on few occasions but apparently has a wide geographic range; apart from the single example dredged in British waters (Fig. 27C), it has been found near Calais, in the Mediterranean, the Indian Ocean (Mauritius) and on the coast of Chile. It has mostly been obtained by dredging sandy or shelly sediments at depths of 40–50 m, but occurs beneath intertidal boulders on the Chilean coast.

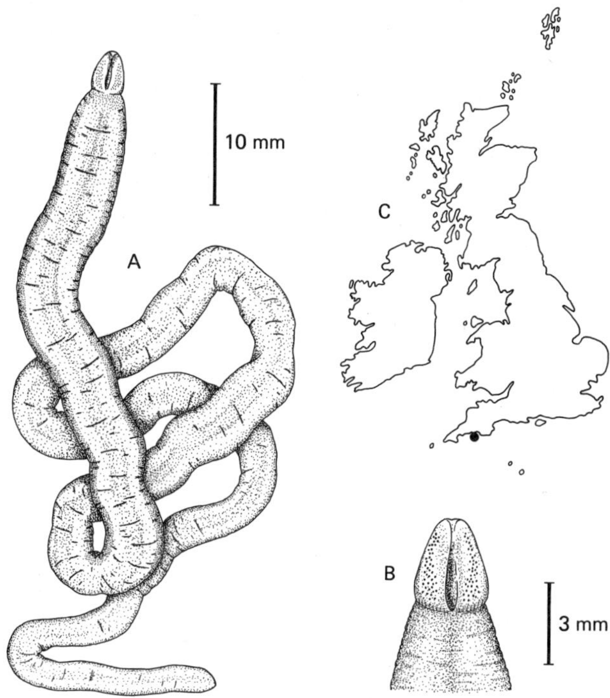

Fig. 27. *Poliopsis lacazei*; A, general view of whole animal; B, dorsal view of cephalic lobe to show the distribution of the eyes and the median dorsal furrow; C, recorded distribution from the British Isles. A, B redrawn from Joubin (1894).

Family VALENCINIIDAE

Genus *VALENCINIA* Quatrefages, 1846

Diagnosis

Heteronemertea without cephalic furrows; proboscis pore ventral, subterminal; proboscis with outer longitudinal and inner circular muscle layers, without muscle crosses; rhynchocoel wall circular muscles not interwoven with body wall musculature; body wall without diagonal muscle layer; dorsal cerebral ganglia posteriorly forked into long upper and lower branches; nervous system with neither neurochords nor neurochord cells; dermis without separate glandular and connective tissue layers; cerebral sensory organs small, not enclosed by blood vessels; blood vascular system in head developed into a cephalic plexus; caudal cirrus absent; cephalic glands very well developed and posteriorly extending into foregut region; excretory system located alongside and below foregut, with lateral nephridiopores; eyes absent; sexes separate.

Valencinia longirostris Quatrefages, 1846
(Fig. 28)

Valencinia longirostris Quatrefages, 1846
Valencia longirostris Koehler, 1885

Specific internal characters

The genus *Valencinia* is monospecific; the specific internal characters are thus as given in the generic diagnosis.

Description

Valencinia longirostris (Fig. 28A) possesses a rather cylindrical body, up to about 15 cm long and 2–3 mm in diameter. The head is slender and sharply pointed, generally coloured white or greyish-white. Most of the body, which is thickest in its posterior half, is pink, yellowish-grey, cinnabar brown or chocolate brown in colour.

This is an English Channel (Fig. 28B) and Mediterranean species, typically found at depths of 1–10 m among the roots of the sea-grasses *Zostera* and *Posidonia* or on sandy sediments. It is doubtful that *Valencinia lineformis* McIntosh, 1873–74 (see p. 189), found on the Scottish coast and regarded as synonymous with *Valencinia longirostris* by Bürger (1895, 1904), represents the same species.

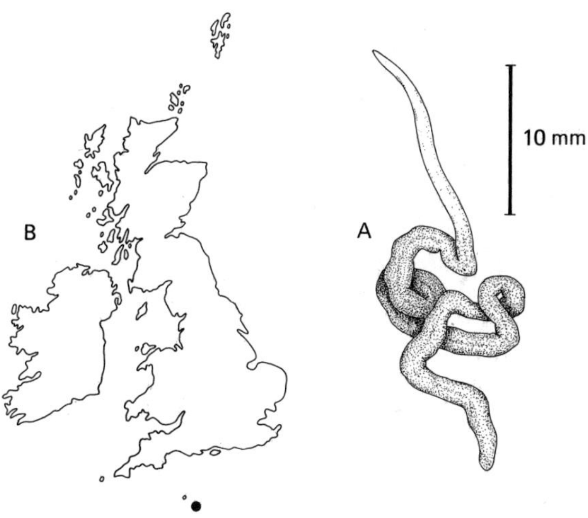

Fig. 28. *Valencinia longirostris*; A, general view of whole animal; B, recorded distribution from the British Isles. A redrawn from Joubin (1894) and Bürger (1895).

Family AMPHIPORIDAE

Genus *AMPHIPORUS* Ehrenberg, 1831

Diagnosis

Monostiliferous hoplonemerteans with usually two pairs of transverse cephalic furrows; rhynchocoel extends to, or almost to, posterior end of body, with wall composed of separate circular and longitudinal muscle layers; proboscis generally well developed; dermis development variably thick or thin; body wall musculature generally well developed, with or without diagonal muscle layer between circular and longitudinal muscles; longitudinal muscle stratum in anterior body regions either a single layer or, less commonly, divided into two in some species by parenchymatous wedge; dorsoventral muscles present but usually few in number and restricted to intestinal regions; nervous system with neither neurochords nor neurochord cells; frontal organ present; cephalic glands usually few in number and small; cerebral sensory organs anterior to cerebral ganglia, variably developed according to species; blood vascular system with three longitudinal vessels, mid-dorsal vessel with single vascular plug; excretory system positioned between rear of cerebral ganglia and foregut; intestinal caecum present, with a pair of anterior diverticula as well as lateral pouches; eyes normally numerous and distributed in groups, occasionally few or none; sexes separate.

Amphiporus allucens Bürger, 1895
(Fig. 29A)

Amphiporus allucens Bürger, 1895; Wijnhoff, 1912; M.B.A., 1931, 1957

Specific internal characters

Not known; there is no information available concerning the internal anatomy of this species.

Description

Because this species is inadequately described neither its generic affinities nor specific validity can be regarded as certain. Even the original designation of the form is confusing; in his text Bürger (1895) merely describes *Amphiporus allucens* as a variety of *Amphiporus pulcher* (now *Nipponnemertes pulcher*: see p. 126, yet in the caption to his illustration records it as a new species. Wijnhoff (1912), in her record of the two British examples found, regards the form as having a specific status.

Amphiporus allucens (Fig. 29A) is up to about 4.0–4.5 cm long and 2.5–3.0 mm wide. The head and anterior trunk are pale yellowish in colour with the cerebral ganglia showing pale pink, the bulk of the body is a bright salmon red hue. On the bluntly pointed head the large, black eyes are arranged in double continuous rows on either side, extending from the tip of the head back to just in front of the cerebral ganglia. *Amphiporus allucens*

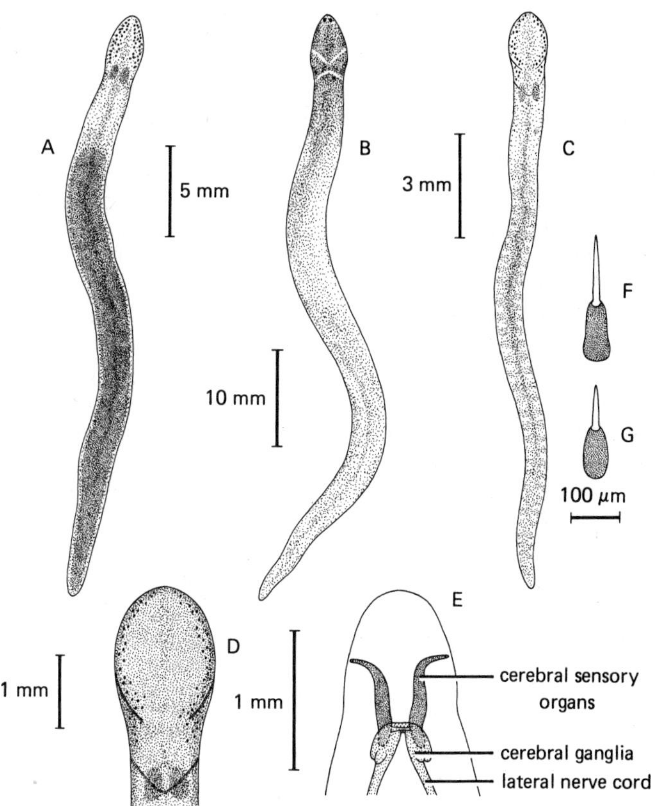

Fig. 29. *Amphiporus allucens*; A, dorsal view of whole animal. *Amphiporus bioculatus*; B, dorsal view of whole animal. *Amphiporus dissimulans*; C, dorsal view of whole animal; D, enlargement of cephalic region to show the distribution of the eyes and position of the cephalic furrows; E, schematic representation of the anterior regions to show the relative positions of the cerebral ganglia and cerebral sensory organs; F, G, variations in size and shape of the central stylet and basis. A redrawn from Bürger (1895), B redrawn from McIntosh (1873–74), D–G redrawn from Berg (1972b).

has only a single pair of transverse cephalic furrows which meet medially on the ventral surface.

Apart from the British records (Fig. 31A), dredged from 20–30 m or more depth, the species is known only from Naples.

Amphiporus bioculatus McIntosh, 1873–74
(Fig. 29B)

Amphiporus bioculatus McIntosh, 1873–74; Riches, 1893; Sheldon, 1896; M.B.A., 1904, 1931, 1957; Southern, 1913; Farran, 1915

Specific internal characters

Body wall longitudinal muscle layer anteriorly divided into inner and outer zones by wedge of parenchymatous connective tissue.

Description

Amphiporus bioculatus (Fig. 29B) attains a length of 8–10 cm and width of 5–6 mm. The heart-shaped to bluntly pointed head bears at its tip a single pair of large eyes. Behind these the cephalic furrows are evident on the dorsal surface, although descriptions of their arrangement are at variance with each other. McIntosh (1873–74), for example, describes but a single pair of furrows, yet in two illustrations depicts two pairs with opposing configurations, one shown as $\genfrac{}{}{0pt}{}{\vee}{\wedge}$, the other as $\genfrac{}{}{0pt}{}{\wedge}{\vee}$. Other authors have described only a single dorsally V-shaped pair of furrows with the apex of the V directed either anteriorly or posteriorly, and a few records indicate that no furrows at all could be distinguished. The body itself is rounded and either gradually increases its diameter through the anterior two-thirds of its length, or is for most of its length of a more or less uniform width. The tail varies from pointed to bluntly rounded.

McIntosh (1873–74) described the colour of *Amphiporus bioculatus* as dull orange or pale brownish, tending towards reddish anteriorly and especially on the head, with a paler ventral surface. Other reports describe the colour as milk white or creamy-white to rose-red, and dark green examples, supposedly of this species, have been recorded from the Channel coast of France.

Friedrich (1955), in a review of the monostiliferous hoplonemertean genera, distinguished an '*Amphiporus hastatus* McIntosh, 1873–74' group from other *Amphiporus* species on the grounds that in this group the body wall longitudinal musculature is anteriorly divided into two by parenchymatous tissues. He included *Amphiporus bioculatus* in the '*hastatus*' group, but Kirsteuer (1974), however, points out that because of differences in the organisation of the blood system the *Amphiporus bioculatus* of McIntosh must be excluded from the '*hastatus*' group. Earlier (Kirsteuer, 1967b), he had separated the Brazilian *Amphiporus bioculatus* and placed them in the genus *Correanemertes*, at the same time commenting that European and North American examples of *Amphiporus bioculatus* might belong to the genus *Paramphiporus*. The problem lies in that different descriptions of *Amphiporus bioculatus* contain conflicting information and, indeed, may relate to different species. Until such time as these taxonomic problems are resolved *Amphiporus bioculatus* must be retained within the genus *Amphiporus*, although subsequent studies may well place it in another monostiliferous group.

Amphiporus bioculatus occurs in sand or among *Laminaria* roots from the lower shore to 30 m or more depth. Mature males have been recorded in August. Because of the taxonomic confusion surrounding records of the species its zoogeographic range is uncertain, but it is reported from the eastern coast of North America, the British Isles (Fig. 31B) and the Channel coast of France at Roscoff.

Amphiporus dissimulans Riches, 1893
(Fig. 29C–G)

Amphiporus dissimulans Riches, 1893; Beaumont, 1895a, b, 1900a, b; Gamble, 1896; Herdman, 1896, 1900; Sheldon, 1896; Jameson, 1898; M.B.A., 1904, 1931, 1957; Southern, 1908a, 1913; Wijnhoff, 1912; Moore, 1937; Bruce *et al.*, 1963

Specific internal characters

Dorsal fibre core of cerebral ganglia posteriorly forked into upper and lower branches; cerebral sensory organs long and slender, projecting well below cerebral ganglia; proboscis usually with 10, occasionally 12, nerves.

Description

After early disagreement about the validity of this species (Wijnhoff, 1912), *Amphiporus dissimulans* has recently been revalidated and fully described by Berg (1972b).

This species (Fig. 29C) reaches lengths of 5.0–7.5 cm and is normally 1–2 mm wide. The body is round, slender and gradually tapers towards the tail. The head is distinct, oval in shape and consistently bears a large number of eyes arranged in a continuous row on each dorsolateral margin. Two pairs of transverse cephalic furrows are present, the posterior pair, which meet mid-dorsally to form a rearwards pointing V, being located above the cerebral ganglia.

The colour of *Amphiporus dissimulans* varies from a light or dark salmon pink to orange, yellowish-brown or light reddish-brown, the intestinal regions often appearing deeper in hue due to the gut contents. The cerebral ganglia are visible as two dark pink or orange lobes a short distance behind the eyes.

Amphiporus dissimulans may be confused with another British species, *Amphiporus lactifloreus*, on external features, although consistently different characters can be recognised.

The species, which has been reported as breeding during the spring and in October, occurs on shelly ground, among ascidians and other organisms on firm mud, under stones on muddy sediments or in sand or gravel from the lower shore to 40 m or more depth. Its geographic range extends from the British Isles (Fig. 31C) to Scandinavia.

Amphiporus elongatus Stephenson, 1911
(Fig. 30A, B, C)

Amphiporus elongatus Stephenson, 1911; King, 1911

Specific internal characters

Cephalic glands poorly developed; proboscis very slender.

Description

A poorly described species so far known from only a single specimen. The body is about 7.5 cm long but very slender and filiform (Fig. 30A), less than 1 mm wide when extended and markedly tapering towards the flattened and bluntly pointed head. The tail is rounded. In colour *Amphiporus elongatus* is a bright yellow on both dorsal and ventral surfaces, but rather more whitish on its lateral margins. When the animal is contracted the colour assumes an orange tinge. As in many other species of *Amphiporus*, the cerebral ganglia show as a pair of reddish lobes at the rear of the head. Two pairs of posteriorly angled transverse cephalic furrows are evident on the dorsal surface of the head; both pairs extend ventrally and anteriorly towards the mid-line. There are five eyes in the only known specimen. The anterior pair, situated near the tip of the head, are some distance from the remaining three eyes, which are located over the large cerebral sensory organs a short way in front of the cerebral ganglia.

Amphiporus elongatus was found in sand and is at present known only from the one locality in the Clyde (Fig. 31A).

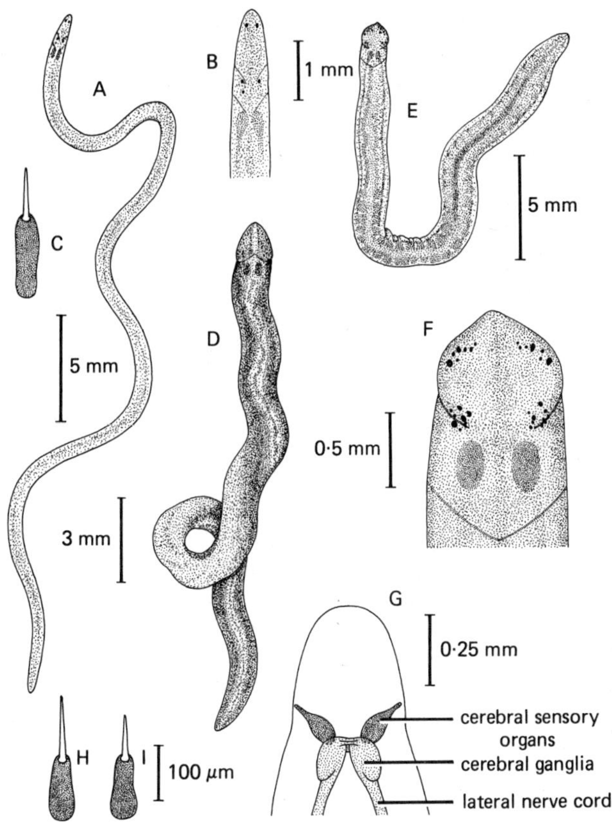

Fig. 30. *Amphiporus elongatus*; A, dorsal view of whole animal; B, enlargement of head to show the eye and cephalic furrow pattern; C, central stylet and basis. *Amphiporus hastatus*; D, general view of complete specimen. *Amphiporus lactifloreus*; E, dorsal view of whole animal; F, enlargement of cephalic region to show the distribution of the eyes and position of the cephalic furrows; G, schematic representation of the anterior regions to show the relative positions of the cerebral ganglia and cerebral sensory organs; H, I, variations in the size and shape of the central stylet and basis. A–C redrawn from Stephenson (1911), D redrawn from McIntosh (1873–74) and Bürger (1895), F–I redrawn from Berg (1972b).

Amphiporus hastatus McIntosh, 1873–74
(Fig. 30D)

Amphiporus hastatus McIntosh, 1873–74, 1875a, b, 1906b, 1927; Riches, 1893; Sheldon, 1896; Southern, 1913; Farran, 1915; Southward, 1953; Bruce *et al.*, 1963; Laverack & Blackler, 1974

Specific internal characters

Body wall longitudinal musculature anteriorly divided into two layers by a wedge of parenchymatous connective tissues; cephalic blood system developed into a lacunar complex.

Description

Amphiporus hastatus is an inadequately described species and, although for the time being retained within the genus *Amphiporus*, certainly belongs with a different group (Friedrich, 1955; Kirsteuer, 1974).

Up to about 9–10 cm or more long when extended, *Amphiporus hastatus* (Fig. 30D) is usually some 3–4 mm in width, although individuals as wide as 15–20 mm are known (M. S. Laverack, personal communication). The head is a bluntly pointed triangular shape and bears numerous eyes; these are described as small and indistinct by McIntosh (1873–74) but large and obvious by Joubin (1894). There is only a single pair of cephalic furrows, both dorsally and ventrally forming a forward pointing V-shape. On the dorsal surface of the head a pale coloured longitudinal median ridge is usually evident. The colour of *Amphiporus hastatus* is variously described as pinkish, yellowish-brown, bright red, pale brown or dark greyish-brown, with the ventral surface and head paler than the remaining body regions. Darker longitudinal pigment patches may occur on the head, and there may also be whitish pigment flecks contributing to the paler cephalic appearance. The cerebral ganglia show distinctly reddish. The proboscis and rhynchocoel are usually evident through the body wall.

Although Southward (1953) states that *Amphiporus hastatus* appears to be intolerant of coarser sediments, it is usually found in sand from mid-tidal level to depths of 35 m or more. Sand grains may adhere to the copious amount of transparent mucus which the nemerteans secrete, forming a fairly firm tube. The geographic range of the species extends from the Mediterranean to the British Isles (Fig. 31A) and Scandinavia; it is also recorded from Greenland and the eastern coast of North America.

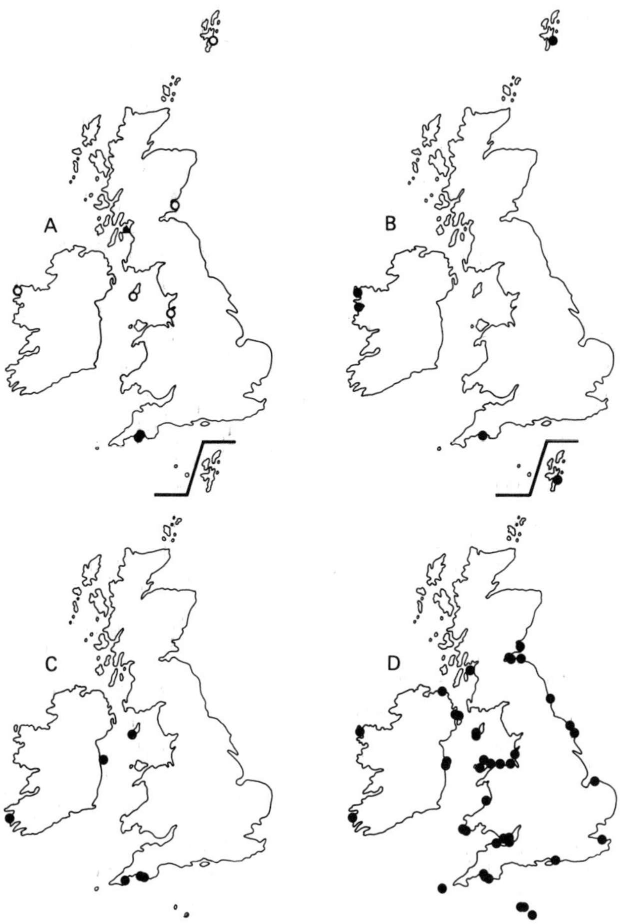

Fig. 31. A, recorded distribution of *Amphiporus allucens* (●), *Amphiporus elongatus* (▲) and *Amphiporus hastatus* (○) from the British Isles; B, recorded distribution of *Amphiporus bioculatus* from the British Isles; C, recorded distribution of *Amphiporus dissimulans* from the British Isles; D, recorded distribution of *Amphiporus lactifloreus* from the British Isles.

Amphiporus lactifloreus (Johnston, 1827–28)
(Fig. 30E–I)

Planaria lactiflorea Johnston, 1827–28
Nemertes Nemertes lactiflorea Johnston, 1837
Nemertes lactiflorea Thompson, 1841
Borlasia alba Thompson, 1843, 1845, 1856; Johnston, 1846
Prostoma lactiflorea Thompson, 1843, 1856; Johnston, 1846
Planaria rosea Thompson, 1845, 1856
Prostoma? rosea Johnston, 1846
Gordius albicans Dalyell, 1853
Omatoplea alba Johnston, 1865
Omatoplea rosea Johnston, 1865
Ommatoplea rosea Lankester, 1866; McIntosh, 1867a
Ommatoplea alba Lankester, 1866; McIntosh, 1867a, 1868a, 1869
Ommatoplea alba var. *rosea* McIntosh, 1869
Amphiporus lactifloreus McIntosh, 1873–74, 1875a, c, 1906b; Koehler, 1885; Haddon, 1886a; Riches, 1893; Vanstone & Beaumont, 1894, 1895; Herdman, 1894, 1900; Beaumont, 1895a, b, 1900a, b; Gamble, 1896; Sheldon, 1896; Jameson, 1898; Gemmill, 1901; Newbigin, 1901; M.B.A., 1904, 1931, 1957; Southern, 1908a, 1913; Evans, 1909; Stephenson, 1911; Wijnhoff, 1912; Walton, 1913; Farran, 1915; Elmhirst, 1922; Moore, 1937; King, 1939; Purchon, 1948, 1956; Cowey, 1952; Eales, 1952; Southward, 1953; Newell, 1954; Bassindale & Barrett, 1957; Barrett & Yonge, 1958; Clark & Cowey, 1958; Bruce *et al.*, 1963; Crothers, 1966; Gibson & Jennings, 1967; Gibson, 1968a; Jennings & Gibson, 1969; Ling, 1971; Williams, 1972; Laverack & Blackler, 1974; Slinger & Gibson, 1974, 1975; Slinger, 1975; Campbell, 1976; Boyden *et al.*, 1977; Withers & Thorp, 1977; Varndell, 1980a, b, 1981a, b
? *Polystemma alba* Gemmill, 1901

Specific internal characters

Dorsal fibre core of cerebral ganglia not distinctly forked at rear; cerebral sensory organs flask-shaped, barely reaching below cerebral ganglia; proboscis usually with 14, sometimes 11–13, proboscis nerves.

Description

One of the commonest British nemertean species, *Amphiporus lactifloreus* (Fig. 30E) has been redescribed by Berg (1972b). Adult individuals 25–35 mm long and 1.0–1.5 mm wide during normal creeping movement are regularly encountered, although McIntosh (1873–74) states that they may occasionally reach a length of 10 cm, and Bürger (1895) found that deep-water examples from near Capri were commonly 10–12 cm long and 2–3 mm wide. The body is rounded, slender and gradually tapers towards the bluntly pointed tail. The head is a rounded oval to spatulate shape, sometimes anteriorly rather pointed, with two pairs of transverse cephalic furrows. In contrast to *Amphiporus dissimulans* (p. 115), with which *Amphiporus lactifloreus* may be confused, the posterior pair of furrows is located behind the cerebral ganglia. Along both sides of the head eyes are distributed in anterior and posterior groups. The number of eyes increases with the age of the animals, but there are never as many as in *Amphiporus dissimulans*, and they do not form a continuous row.

Amphiporus lactifloreus is quite variable in colour, depending upon its reproductive state and the amount of food in the gut. In general it is a dull pinkish or dirty white hue, with paler head, tail and body margins. The cerebral ganglia are visible dorsally as dark pink or reddish patches. During the reproductive period (April to June) the eggs of mature females may impart an orange to rich reddish colouration to the intestinal regions, whereas males appear a creamish-grey to light brownish-grey tint. Gut contents at all times of the year may give the post-cephalic body regions a dark grey, muddy-brown, greenish, or, occasionally, almost black colour. Specimens in which the head appears dark brown to black are likely to be infected with the sporozoan parasite *Haplosporidium malacobdellae* (Varndell, 1980b, 1981a, b).

Widely distributed in the British Isles (Fig. 31D) and locally very abundant on fairly clean sandy or gravelly sediments from just below the *Pelvetia* zone on the shore to depths of 250 m or more, *Amphiporus lactifloreus* is also found amongst shell debris, on *Laminaria*, *Fucus* and *Ascophyllum* and, less commonly, in silty or muddy substrata. It is to some extent tolerant of fluctuating salinities.

The geographic distribution of *Amphiporus lactifloreus* extends from the Mediterranean and northern coasts of Europe to the Atlantic and Arctic coasts of North America north of Cape Code.

Family CARCINONEMERTIDAE

Genus *CARCINONEMERTES* Coe, 1902

Diagnosis

Monostiliferous hoplonemerteans ectohabitant on the gills or egg masses of crabs; without cephalic furrows; eyes two, rarely with one or both subdivided by fragmentation; rhynchocoel short, reaching but a little distance behind cerebral ganglia and without muscular walls, consisting only of thin lining membrane closely applied to proboscis; proboscis reduced, small, very short, with central stylet and basis but without accessory stylets and pouches, anterior proboscis with non-glandular epithelium; dermis thin; body wall musculature weakly developed; frontal organ absent; cephalic glands very well developed, opening to exterior via improvised ducts; submuscular glands extremely well developed throughout body; cerebral sensory organs absent; nervous system with neither neurochords nor neurochord cells; blood vascular system with two longitudinal vessels only, without transverse connectives other than at front and rear; excretory system absent; intestinal caecum absent, stomach opening directly into anterior of intestine, intestine with lateral diverticula; sexes separate, males with testes discharging into single internal duct (Takakura's duct) which in turn opens into rear of intestine, usually oviporous, although internal fertilisation with subsequent development to free-swimming embryos may occur infrequently.

Carcinonemertes carcinophila (Kölliker, 1845)
(Fig. 32)

Nemertes cartinophilos Kölliker, 1845
Polia involuta McIntosh, 1869
Nemertes carcinophilus McIntosh, 1869
Nemertes carcinophila McIntosh, 1873–74, 1875c, 1927; Riches, 1893; Sheldon, 1896; Gemmill, 1901
Carcinonemertes carcinophila Wijnhoff, 1912; M.B.A., 1931, 1957; Laverack & Blackler, 1974

Specific internal characters

Anal blood vessel commissure passes ventrally below posterior end of intestine; ovaries arranged in single row on each side of intestine; diameter of central stylet basis 6–8 μm.

Description

There are two varieties of this species (Humes, 1942); the British form is *Carcinonemertes carcinophila carcinophila* (Fig. 32A). Sexually mature adults, which live on the egg masses of crabs, are 20–70 mm long, whereas juvenile worms, found on the host gills, are only about 15 mm in maximum length. The body is slender and yellowish, orange, pale reddish, rose pink or bright brick red in colour. Two small eyes are evident near the anterior tip,

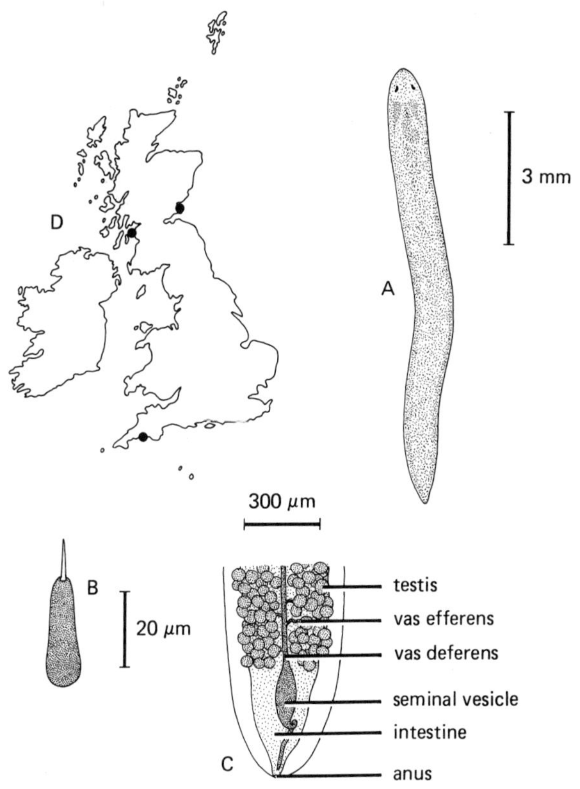

Fig. 32. *Carcinonemertes carcinophila*; A, dorsal view of whole animal; B, a central stylet and basis; C, schematic representation of the posterior region of a mature male to show the organisation of the reproductive system; D, recorded distribution from the British Isles. B, C redrawn from Humes (1942).

although occasionally there may appear to be more because one or both eyes have become fragmented.

So far as is known the entire life cycle of *Carcinonemertes* is spent on the crab hosts, although artificially removed specimens are capable of surviving for several weeks in clean sea water. Juvenile worms live among their host gill filaments, adhering to them by means of sticky mucous secretions. In a heavy infestation some of the gill plates may become damaged, blackened and degenerated. The nemerteans remain on the gills until the host produces eggs, then migrating to the egg masses where they attain sexual maturity and lay their own eggs in mucous tubes. The newly hatched nemerteans tend to remain among the host eggs until the end of the breeding period, after which they move to the gill chamber.

Carcinonemertes carcinophila carcinophila has been found on galatheid, portunid and xanthid crabs from Europe to the Atlantic coast of North America. In the British Isles recorded hosts are the portunids *Carcinus maenas* (Linnaeus) and *Liocarcinus depurator* (Linnaeus). 40–50 or more nemerteans may be found upon a single host. The related variety, *Carcinonemertes carcinophila imminuta*, which differs from the British form in length and in the size of the proboscis armature, is found in the Gulf of Mexico, the West Indies, Central America and Brazil.

Family CRATENEMERTIDAE

Genus *NIPPONNEMERTES* Friedrich, 1968

Diagnosis

Monostiliferous hoplonemerteans with two pairs of transverse cephalic furrows; mostly with numerous large eyes, but one species eyeless; rhynchocoel full body length, with wall composed of a wickerwork of interwoven longitudinal and circular muscle fibres; proboscis moderately well developed; dermis thin; body wall musculature, especially longitudinal layer, well developed; except for one species, without diagonal muscle layer between body wall circular and longitudinal muscles; frontal organ present; cephalic glands usually well developed and posteriorly reaching cerebral ganglia; cerebral sensory organs large, primarily situated posterior to cerebral ganglia; nervous system with neither neurochords nor neurochord cells, without accessory lateral nerve; blood vascular system with three longitudinal vessels, mid-dorsal vessel with single vascular plug; excretory system well developed, extending from rear of cerebral ganglia to hind regions of foregut; intestinal caecum without anterior diverticula but with lateral pouches; sexes separate.

Nipponnemertes pulcher (Johnston, 1837)
(Fig. 33)

Nemertes Nemertes pulchra Johnston, 1837
Prostoma pulchra Johnston, 1846
Vermiculus rubens Dalyell, 1853
Omatoplea pulchra Johnston, 1865
Ommatoplea pulchra Lankester, 1866; McIntosh, 1868a, b, 1869
Amphiporus pulcher McIntosh, 1873–74, 1875b, c; Haddon, 1886b; Riches, 1893; Herdman, 1894, 1896, 1900; Vanstone & Beaumont, 1894, 1895; Beaumont, 1895a, b, 1900b; Sheldon, 1896; Jameson, 1898; Allen, 1899; Gemmill, 1901; M.B.A., 1904, 1931, 1957; Stephenson, 1911; Southern, 1913; Elmhirst, 1922; Moore, 1937; Jones, 1939, 1951; Bruce *et al.*, 1963; Laverack & Blackler, 1974

Specific internal characters

No diagonal layer in body wall musculature; proboscis usually with 12, less often 8–11 or 13–14 nerves.

Description

Transferred to the genus *Nipponnemertes* by Berg (1972a), *Nipponnemertes pulcher* (Fig. 33A) is up to 9 cm long and 1–5 mm broad, with a rather stout, somewhat dorsoventrally compressed, body which gradually tapers posteriorly to the bluntly pointed tail. The head is distinct and shaped like a rounded triangle or shield. It bears two pairs of cephalic furrows. The anterior pair, dorsally incomplete, possesses short forwardly-directed parallel ridges running perpendicular to the transverse furrows. A median longitudinal ridge or swelling runs from the tip of the head to the most anterior part of the body, where the posterior cephalic furrows meet to form a wide V-shape with the apex pointing caudally. The posterior furrows do not reach the mid-line on the ventral surface and are not as distinct as the anterior pair, especially in pale coloured worms.

The eyes are distributed irregularly near the lateral cephalic margins, mostly in front of the anterior cephalic furrows. Their number increases with age; Berg (1972a) found that in young individuals of less than 10 mm length there were 10–20 large eyes, in contrast to older specimens 40–50 mm long which had 70–80 smaller eyes. Irrespective of eye number, their arrangement is remarkably constant.

Dorsally *Nipponnemertes pulcher* is coloured brown, red or pink, but the lateral and ventral surfaces are always much lighter and may even appear completely unpigmented. The dorsal longitudinal ridge on the head is sometimes darker than the rest of the body, and the ridge is emphasised through the absence of pigmentation from its lateral margins. Body colouration is affected by age, degree of sexual maturation and habitat. Juveniles and mature males are lighter in hue than the dark red colour of the ovaries in gravid females, and individuals maintained in an aquarium for some time

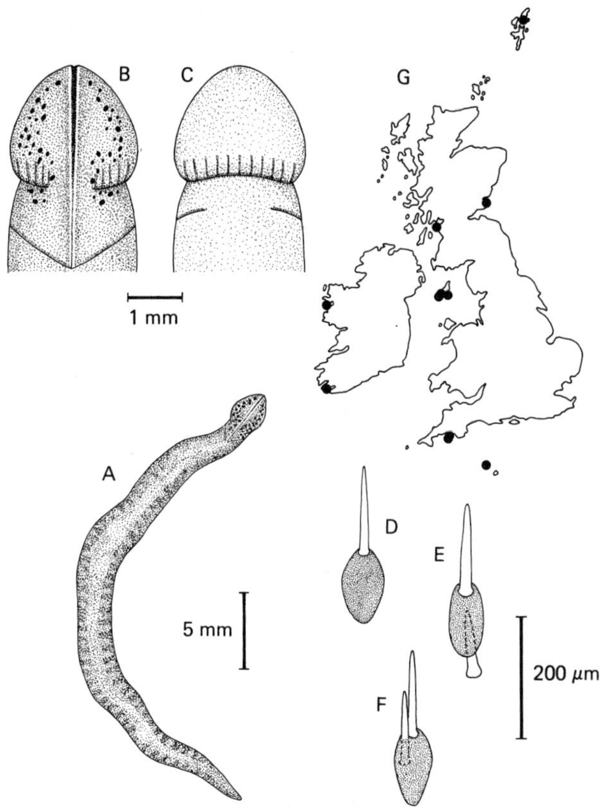

Fig. 33. *Nipponnemertes pulcher*; A, dorsal view of whole animal; B, C, enlargements of the cephalic lobe and anterior body region in dorsal, B, and ventral, C, aspect to show the distribution of the eyes, the median dorsal longitudinal ridge and the arrangement of the cephalic furrows; D–F, variations in the size and shape of the central stylet and basis; in E and F prematurely liberated accessory stylets have become embedded in the central stylet basis; G, recorded distribution from the British Isles. A redrawn from Bürger (1895), B–F redrawn from Berg (1972a).

tend to develop a deeper colouration. Ova develop at the beginning of May and are ripe towards the end of June.

The accessory stylets are often liberated prematurely from the proboscis in this species, with the result that additional stylets are quite often found embedded in the stylet basis (Berg, 1972a).

Around the British Isles *Nipponnemertes pulcher* has been found sublittorally among corallines or on coarser sediments such as sand, gravel or shelly debris at depths down to 240m. In other areas it is commonly found on muddy or stony bottoms and as deep as 569 m. Very occasionally it may occur beneath stones at the extreme lower levels of the shore.

Nipponnemertes pulcher has a wide geographic range. In the northern hemisphere it is reported from the east coast of North America, Greenland, the Faroe Islands, the White Sea and northern Europe from the Atlantic coast of France to Scandinavia. In the southern hemisphere it is known from Chile and many parts of the Antarctic and sub-Antarctic areas. Berg (1972a) believes that records of the species from the Mediterranean are of doubtful validity.

Family EMPLECTONEMATIDAE

Genus *EMPLECTONEMA* Stimpson, 1857

Diagnosis

Monostiliferous hoplonemerteans with transverse but often indistinct cephalic furrows; with numerous small eyes; rhynchocoel restricted to anterior half of body, with wall composed of separate circular and longitudinal muscle layers; proboscis short, slender; dermis variably developed; body wall musculature, especially longitudinal layer, well developed but without diagonal zone; strong dorsoventral musculature present in head; frontal organ present or absent; cephalic glands well developed, sometimes reaching to or beyond cerebral ganglia; cerebral sensory organs either large and close to anterior margins of cerebral ganglia or small and far in front of brain lobes; nervous system with neither neurochords nor neurochord cells, without accessory lateral nerve; blood vascular system with three longitudinal vessels, mid-dorsal vessel with single vascular plug; excretory system located in region between mid-brain and foregut; intestinal caecum absent or very short, but two anterolateral diverticula present; sexes separate, gonads small and numerous.

Emplectonema echinoderma (Marion, 1873)
(Fig. 34)

Borlasia echinoderma Marion, 1873
Nemertes echinoderma Punnett, 1901a
Eunemertes echinoderma M.B.A., 1904
Emplectoneema echinoderma Southern, 1913
Emplectonema echinoderma Wijnhoff, 1912; Farran, 1915; M.B.A., 1931, 1957

Specific internal characters

Frontal organ present; cephalic glands not extending to cerebral ganglia; intestinal caecum present; cerebral sensory organs large, close to anterior margins of cerebral ganglia.

Description

A rare British species, *Emplectonema echinoderma* (Fig. 34A) is easily identified by the presence of large numbers of minute transparent and sickle-shaped spicules scattered throughout its epidermis (Fig. 34D). Up to 20 cm long, the body is 1.0–1.5 mm wide anteriorly, but up to 2.5 mm across in the intestinal regions. The hind end gradually narrows to terminate in a bluntly pointed tail.

The head is distinct and possesses a rounded diamond shape. It bears a row of about 20 small eyes on each side which extend from the tip back to the cerebral ganglia. *Emplectonema echinoderma* is an overall pale salmon, yellowish-red or orange-red colour, generally paler in the posterior third of the body. The head may be dorsally marked with darker longitudinal patches of pigmentation. According to Bürger (1895) small individuals may be white or colourless.

Also known from Madeira and the Mediterranean, *Emplectonema echinoderma* in the British Isles has been found in lower-shore and shallow water situations, burrowed in sand, beneath stones or in *Zostera* beds. On one occasion a number of individuals was found 'commensal with *Leptosynapta inhaerens*' (a holothurian echinoderm) (M.B.A., 1931, 1957), but it is doubtful that the nemertean is an habitual commensal.

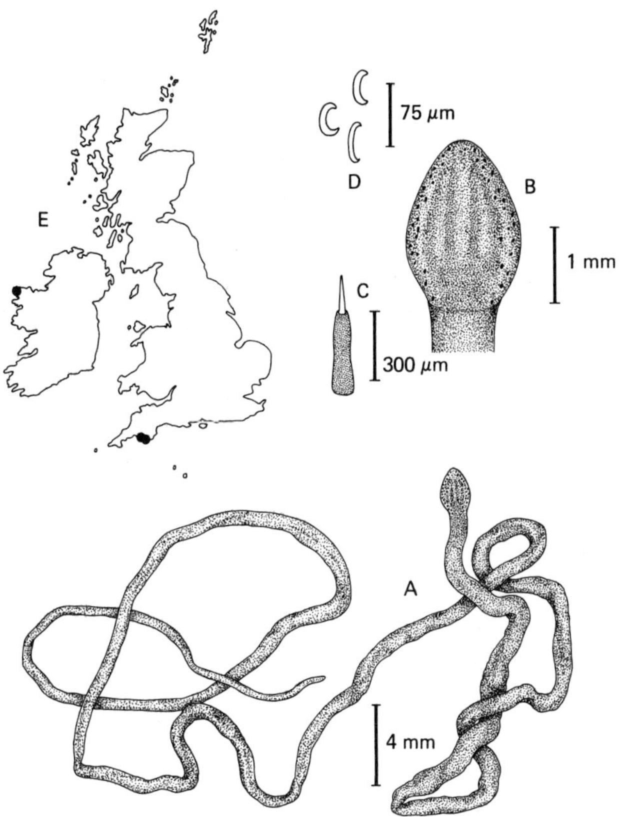

Fig. 34, *Emplectonema echinoderma*; A, general view of complete individual; B, enlargement of head to show the distribution of the eyes; C, a central stylet and basis; D, a group of epidermal spicules; E, recorded distribution from the British Isles. A, B, redrawn from Bürger (1895), C redrawn from Marion (1873), D redrawn from Joubin (1894).

Emplectonema gracile (Johnston, 1837)
(Fig. 35)

Nemertes Nemertes gracilis Johnston, 1837
Nemertes gracilis Thompson, 1841; McIntosh, 1873–74, 1875b, c; Riches, 1893; Sheldon, 1896
Prostoma gracilis Thompson, 1843, 1856; Johnston, 1846
Omatoplea gracilis Johnston, 1865
Ommatoplea gracilis McIntosh, 1868a, 1869
Eunemertes gracilis Jameson, 1898; M.B.A., 1904; Elmhirst, 1922; Eales, 1952
Emplectoneema gracile Southern, 1913
Eunemertes gracile Barrett & Yonge, 1958
Emplectonema gracile King, 1911; Stephenson, 1911; Wijnhoff, 1912; M.B.A., 1931, 1957; Bruce, 1948; Bassindale & Barrett, 1957; Bruce *et al.*, 1963; Crothers, 1966; Eason, 1973; Laverack & Blackler, 1974; Slinger & Gibson, 1974, 1975; Slinger, 1975

Specific internal characters

Cerebral sensory organs small, located some distance in front of cerebral ganglia; cephalic glands voluminous, posteriorly extending into foregut regions.

Description

A fairly common species on rocky shores, *Emplectonema gracile* (Fig. 35A) attains lengths of up to about 50 cm but is only 3–4 mm in maximum width. The head is rounded, wider than the adjacent trunk, rather flattened dorsoventrally and bears 20–30 or more small eyes on each side arranged into antero- and posterolateral groups of about equal numbers. In small and young individuals there are generally fewer eyes and their distribution is irregular. The body is long and slender, very contractile and of a more or less uniform width throughout most of its length, only narrowing posteriorly to end in a blunt rounded tail. The dorsal colouration varies from a dull olive-green to greyish-green or a dark blue-green, occasionally greenish-brown. Often irregularly shaped and distributed small dark green or black pigment flecks are scattered over the dorsal and dorsolateral surfaces. A blue-grey iridescence, due to the flickering of the epidermal cilia, is sometimes apparent. At the rear of the head the cerebral ganglia show as two dull red lobes. Ventrally the body is a pale greyish-yellow, greyish-green or dirty yellowish-white colour, through which mature gonads show white or yellowish. Usually the ventral colouration appears as a distinct broad longitudinal stripe clearly bordered by ventrolateral extensions of the dorsal pigmentation.

Emplectonema gracile reproduces during the months of April, May and June. It occurs intertidally beneath stones and boulders on coarse muddy silt, shelly gravel or silty sand, in rock crevices and clefts, in cavities in boulders,

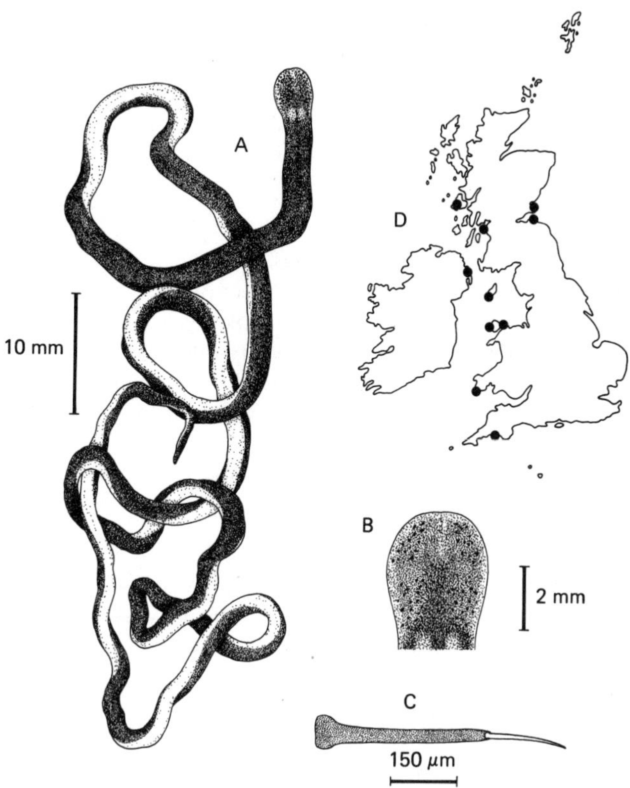

Fig. 35. *Emplectonema gracile*; A, general view of whole animal; B, enlargement of head to show the distribution of the eyes; C, a central stylet and basis; D, recorded distribution from the British Isles.

in mussel beds or among the roots of laminarians. Mostly occupying the mid- to lower-shore levels, the species may occasionally be found crawling between barnacles high up the shore when the tide is out. Copious amounts of a thick and sticky mucus are secreted when the animal is disturbed, and fine particles of mud or silt may adhere to the mucus to form a tube in which the animal rests. Among algal holdfasts or beneath boulders *Emplectonema gracile* is most frequently found with its body in a knotted tangle which is extraordinarily difficult to unravel without rupturing it. Occasionally it is dredged from coarser sediments in shallow water and in other parts of the world has been found at depths down to 100 m.

The species possesses a wide geographic range and is recorded from the west coast of North America, Chile, the Aleutian Islands, the northern coasts of Europe, the Mediterranean, Madeira, the Kamchatka Peninsula of Russia and Japan.

Emplectonema neesii (Örsted, 1843)
(Fig. 36)

? *Gordius maculosus* Montagu, 1808
? *Lineus maculosus* Montagu, 1808
Planaria flaccida Johnston, 1827–28
Amphiporus neesii Örsted, 1843
Borlasia? *flaccida* Johnston, 1846
Gordius fuscus Dalyell, 1853
Serpentaria fusca Johnston, 1865; Lankester, 1866
? *Omatoplea*? *maculosa* Parfitt, 1867
Ommatoplea purpurea McIntosh, 1868a, b, 1869
Nemertes neesii McIntosh, 1873–74, 1875c; Haddon, 1886b; Riches, 1893; Herdman, 1894, 1896, 1900; Vanstone & Beaumont, 1894, 1895; Beaumont, 1895a, b, 1900a; Gamble, 1896; Sheldon, 1896; Gemmill, 1901; Evans, 1909; Chumley, 1918
Eunemertes neesii Jameson, 1898; Eales, 1952; Barrett & Yonge, 1958; Thompson *et al.*, 1966
Nemertes neesi Beaumont, 1900b
Eunemertes neesi M.B.A., 1904
Emplectonema neesi Wijnhoff, 1912; M.B.A., 1931, 1957; Moore, 1937; Corlett, 1947; Bassindale & Barrett, 1957; Bruce *et al.*, 1963; Crothers, 1966; Thompson *et al.*, 1966; Laverack & Blackler, 1974; Boyden *et al.*, 1977
Emplectoneema neesi Southern, 1913
Emplectonema neesii Southern, 1908a; Stephenson, 1911; Farran, 1915; Whitfield, 1972; Eason, 1973

Specific internal characters

Cerebral sensory organs small, far anterior to cerebral ganglia; cephalic glands well developed but barely or not posteriorly extending behind cerebral ganglia.

Description

Mostly less than 40–50 cm long, occasional specimens of over 1 m length have been found. In larger individuals the body may be 5–6 mm wide, with an elongate, rather flattened and generally wrinkled appearance, but even in smaller examples the worms appear rather more bulky in build than *Emplectonema gracile.* The head is rounded and spatulate in shape (Fig. 36B); it may or may not be wider than the adjacent trunk region. The arrangement of the eyes is similar to that of *Emplectonema gracile*, although their larger number generally leads to them appearing more densely clustered.

In colour *Emplectonema neesii* presents a speckled or longitudinally mottled light to dark brown appearance dorsally, with a tendency to be paler on the anterior and posterior extremities. Closer examination reveals that the

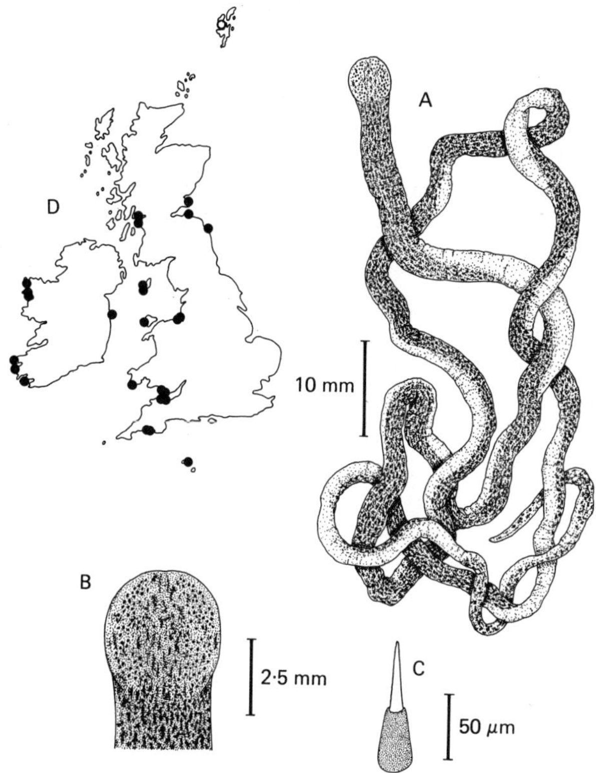

Fig. 36. *Emplectonema neesii*; A, general view of whole animal; B, enlargement of the head to show the distribution of the eyes; C, a central stylet and basis; D, recorded distribution from the British Isles; ● = certain records, ○ = specific identity unconfirmed.

background is a pale yellowish-brown, straw or flesh colour, marked with irregular dark brown pigment specks or streaks which, though loosely arranged in longitudinal rows, are frequently interrupted and do not form distinct stripes. In darker appearing individuals the streaks are often less numerous but more boldly marked than in paler examples. The ventral surface is a pale pinkish-white or flesh colour. Young worms, from deeper water, may on occasion appear almost white or pale skin coloured, and uniformly pigmented dull orange to pale brick-red specimens have also been found.

Emplectonema neesii is found intertidally and sublittorally to 30 m or more depth. On the shore it typically occurs in the mid- to lower-tidal levels beneath stones and boulders, in rock crevices and fissures, in laminarian holdfasts, among the byssus threads of *Mytilus* colonies, or on a variety of substrata (sand, silty-sand, shelly-gravel, shingle). Infrequently it may occur in deep narrow rocky clefts at or just above the *Pelvetia* zone. The species is reproductively active for most of the year, having been reported breeding at any time between January and October.

With a much more restricted zoogeographic range than *Emplectonema gracile*, *Emplectonema neesii* has been recorded only from the Atlantic, Irish Sea and North Sea coasts of Europe, from Iceland to the English Channel, and the Mediterranean.

Genus *NEMERTOPSIS* Bürger, 1895

Diagnosis

Filiform monostiliferous hoplonemerteans with two pairs of distinct transverse cephalic furrows; rhynchocoel at most about half body length, with wall composed of two distinct muscle layers; cephalic dorsoventral muscles feebly developed; cephalic glands moderately well developed; nervous system with neither neurochords nor neurochord cells, without accessory lateral nerve; cerebral sensory organs small, located far anterior to cerebral ganglia; four large or small eyes; blood vascular system with three longitudinal vessels, mid-dorsal vessel with single vascular plug; excretory system present only in foregut region; frontal organ present; intestinal caecum present but short and without anterior diverticula; sexes separate.

Nemertopsis flavida (McIntosh, 1873–74)
(Fig. 37)

Tetrastemma flavida McIntosh, 1873–74, 1875c
Tetrastemma flavidum Riches, 1893; Horsman, 1938
? *Prostoma flavidum* King, 1911
Nemertopsis flavida Beaumont, 1900a, b; M.B.A., 1904, 1931, 1957; Wijnhoff, 1912; Southern, 1913; Farran, 1915; Purchon, 1948; Barrett & Yonge, 1958; Williams, 1972; Laverack & Blackler, 1974
Nemertopsis tenuis Beaumont, 1900a, b; M.B.A., 1904

Specific internal characters

Cephalic glands posteriorly extend behind cerebral ganglia.

Description

Beaumont (1900b), Wijnhoff (1912) and Southern (1913), among others, have inclined to the view that *Nemertopsis flavida* and *Nemertopsis tenuis* Bürger, 1895, are one and the same species. Minor differences between the two forms are reported, including the relative length of the rhynchocoel and presence or absence of a red pigmentation in the blood system, but whether or not these are taxonomically significant is not at present known. Accordingly, both forms are here included as *Nemertopsis flavida*; future studies must determine whether *Nemertopsis tenuis* should be recognised as a separate species.

Nemertopsis flavida (Fig. 37A) reaches a length of some 40 mm but is only about 0.5 mm wide. The filiform body is rather flattened and anteriorly and posteriorly tapered. The bluntly rounded head bears two pairs of rather small black eyes, the anterior pair usually being the largest. The colour is a uniform white, yellowish, pink, pale peach or reddish-brown, with pale lateral margins and a translucent snout. The alimentary tract and rhynchocoel are sometimes visible through the body wall and gut contents may affect the colour in the intestinal regions. The blood system may be indistinguishable in life or be visible because of a red pigmentation; individuals with the latter condition have often been recorded as *Nemertopsis tenuis*.

Nemertopsis flavida is found both intertidally and sublittorally. It occurs in rock pools, beneath stones and boulders, in laminarian holdfasts, among algal fronds, in *Zostera* beds or in a range of sediments including shelly gravel, mud and sand. Sublittorally the species has been dredged from a depth of more than 300 m but is generally recovered from much shallower locations. Sexually mature individuals have been found in May, though in Danish waters specimens with ripe gonads are present during October and November.

The geographic range extends from Denmark to the Mediterranean, although Wheeler (1934) also records it (as *Nemertopsis tenuis*) from South Africa.

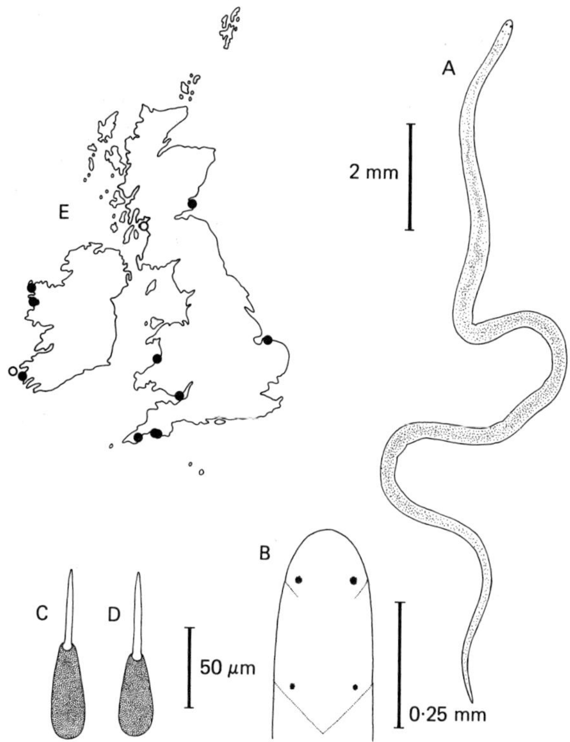

Fig. 37. *Nemertopsis flavida*; A, dorsal view of complete individual; B, enlargement of head to show the four eyes and the pattern of the cephalic furrows; C, D, variations in the size and shape of the central stylet and basis; E, recorded distribution from the British Isles; ● = reasonably certain records, ○ = specific identity uncertain. A redrawn from McIntosh (1873–74) and Bürger (1895), C, D redrawn from McIntosh (1873–74).

Family PROSORHOCHMIDAE

Genus *ARGONEMERTES* Moore & Gibson, 1981

Diagnosis

Monostiliferous terrestrial hoplonemerteans with transverse cephalic furrows; 20–180 eyes, the number increasing with age and size; rhynchocoel full body length, with wall composed of a wickerwork of interwoven longitudinal and circular muscle fibres; proboscis very large, powerfully developed, used for rapid locomotion; dermis thick; body wall musculature usually strongly formed; frontal organ absent; cephalic glands extensive, opening via improvised pores; cerebral sensory organs large, with anterior sac and forked cerebral canal, opening ventrally from transverse cephalic furrows; nervous system with neurochords and neurochord cells, with accessory lateral nerve; blood vascular system with prominent valves but without extra-vascular pouches, developed into an extensive submuscular capillary network, mid-dorsal blood vessel with two vascular plugs; excretory system extensive, with scattered mononucleate flame cells which lack cuticular support bars, excretory tubules with specialised thick-walled terminal region; sexes separate or hermaphroditic.

Argonemertes dendyi (Dakin, 1915)
(Fig. 38)

Geonemertes dendyi Dakin, 1915; Waterston & Quick, 1937; Pantin, 1944, 1947, 1950, 1961, 1969; Durham, 1948, 1949; Hunt, 1948; Danielli & Pantin, 1950; Clark & Cowey, 1958; Cloudsley-Thompson & Sankey, 1961; Jones, 1978

Argonemertes dendyi Anderson, 1980; Moore & Gibson, 1981

Specific internal characters

Body wall musculature only weakly developed; cephalic blood system a capillary network; hermaphroditic.

Description

Argonemertes dendyi (Fig. 38A) is up to about 25 mm long and 1.0–1.5 mm broad; since the species is protandrous and hermaphroditic, males are generally smaller than females. The nemerteans are a cream or pale yellowish background colour with a longitudinal dark brown stripe running along each dorsolateral margin and extending from behind the eyes to the tail. The colouration in older worms is frequently more intense in the posterior half of the body. The stripes may be so broad that only a slender median pale longitudinal line is left between them, in other instances the stripes may be barely distinguishable. The ventral surface is more uniformly paler coloured and lacks stripes. Pantin (1944) records that the background colour may vary from almost white to orange, dark brown or purplish pink.

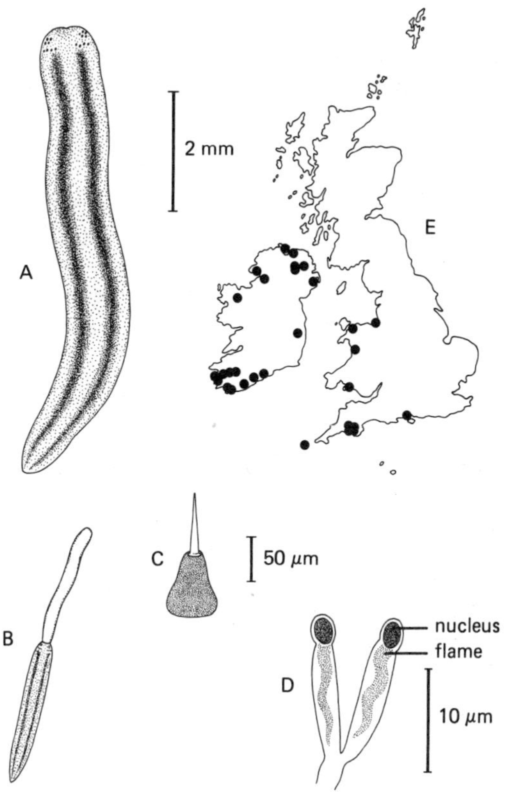

Fig. 38. *Argonemertes dendyi*; A, dorsal view of whole animal; B, an individual with the massive proboscis fully everted; C, a central stylet and basis; D, a pair of flame cells; E, recorded distribution from the British Isles. A, B redrawn from Pantin (1950), D redrawn from Pantin (1969).

The eyes are arranged into four groups with three to eight eyes in each.

In its external features *Argonemertes dendyi* closely resembles terrestrial flatworms such as *Rhynchodemus sylvaticus* (Leidy), with which it may be found, but can be readily distinguished by its characteristic escape behaviour in response to mechanical irritation. When disturbed *Argonemertes dendyi* everts its long white proboscis (Fig. 38B), which flatworms do not possess, grips the substrate with the proboscis tip and then rapidly retracts its body over the organ; in this way the nemertean moves much more rapidly than it can by the more usual means of locomotion, creeping upon a mucoid track.

Eggs encapsulated in a transparent but fairly tough egg mass, about three millimetres long and filled with a thin watery jelly, may be found in damp or sodden moss. From 3–30 eggs may be laid at a time, although 10–15 is the usual number. Reproduction occurs during the period October to April.

Under adverse conditions (e.g. drought) *Argonemertes dendyi* rapidly encapsulates itself in a cocoon of tough mucus. The cocoon, which is very sticky, is attached to the substrate by threads. The nemerteans are able to escape quickly once better conditions prevail.

The species is found in damp and shaded locations, under logs or fallen branches, beneath stones or among decaying leaves and moss. Millipedes, terrestrial flatworms, small oligochaetes, snails and slugs frequent the same types of habitat. The nemerteans have been observed feeding on small nymphs of delphacid bugs, young collembolans and myriapods.

Argonemertes dendyi is a native of south-western Australia, but has been recorded as an immigrant in European greenhouses, the Azores and the Canary Islands as well as the British Isles (Moore & Gibson, 1981).

Several specimens of a terrestrial nemertean, identified as *Geonemertes chalicophora* (Graff, 1879) (now *Leptonemertes chalicophora*: see Moore & Gibson, 1981), were reported from beneath flower pots in hot-houses at the Botanic Gardens, Glasnevin, Dublin, by Southern (1911). *Leptonemertes chalicophora* is indigenous on North Atlantic islands but has been recorded as an immigrant to European greenhouses (Moore & Gibson, 1981). Southern's (1911) specimens, however, were inadequately described and their specific identity must remain uncertain.

Genus *OERSTEDIA* Quatrefages, 1846

Diagnosis

Monostiliferous hoplonemerteans without cephalic furrows; four eyes, the posterior pair sometimes larger than the anterior; rhynchocoel full body length, with wall composed of a thin wickerwork of interwoven circular and longitudinal muscle fibres; proboscis large and well developed but not employed for locomotion; body of a firm consistency, with moderately well developed dermis and body wall musculature but lacking dorsoventral muscles and longitudinal cephalic muscles; frontal organ absent; cephalic glands extensive; cerebral sensory organs small, anterior to eyes; cerebral ganglia small and close-set; nervous system with neither neurochords nor neurochord cells but with accessory lateral nerve; blood vascular system with three longitudinal vessels, without submuscular capillary network, with distinct valves, number of vascular plugs on mid-dorsal vessel uncertain; intestinal caecum present but without anterior diverticula, intestinal diverticula mostly orientated dorsally; sexes separate.

Oerstedia dorsalis (Abildgaard, 1806)
(Fig. 39)

Planaria dorsalis Abildgaard, 1806
? *Planaria gordius* Montagu, 1808
Vermiculus variegatus Dalyell, 1853
Tetrastemma variegatum Johnston, 1865; Lankester, 1866; McIntosh, 1868b, 1869
? *Astemma gordius* Parfitt, 1867
Tetrastemma maculatum? Sumner, 1894
Tetrastemma dorsale Riches, 1893; Herdman, 1894, 1900; Beaumont, 1895a, b, 1900a, b; Vanstone & Beaumont, 1894, 1895; Gamble, 1896; Sheldon, 1896; Allen, 1899
Tetrastemma dorsalis Haddon, 1886a; McIntosh, 1873–74, 1875c; Gemmill, 1901; Laverack & Blackler, 1974
Tetrastemma (Oerstedia) dorsalis Evans, 1915
Oerstedia dorsalis var. *marmorata* Wijnhoff, 1912; M.B.A., 1931, 1957; Bruce *et al.*, 1963
Oerstedia dorsalis var. *viridis* Wijnhoff, 1912; M.B.A., 1931, 1957
Oerstedia dorsalis var. *cincta* Wijnhoff, 1912; M.B.A., 1931, 1957
Oerstedia dorsalis Jameson, 1898; M.B.A., 1904, 1931, 1957; Southern, 1908a, 1913; Stephenson, 1911; Wijnhoff, 1912; Farran, 1915; Moore, 1937; Bruce, 1948; Purchon, 1948; Bassindale & Barrett, 1957; Barrett & Yonge, 1958; Bruce *et al.*, 1963; Crothers, 1966; Thompson *et al.*, 1966; Williams, 1972; Campbell, 1976; Boyden *et al.*, 1977; Knight-Jones & Nelson-Smith, 1977

Specific internal characters

Not certain; it is probable that the name *Oerstedia dorsalis* encompasses a complex of species (P. Sundberg, personal communication).

Description

Oerstedia dorsalis (Fig. 39A) is a small but rather stout species; up to 30 mm long, most examples found are 10–15 mm in length and 1–2 mm wide. The bluntly rounded head, which bears four distinct eyes, is not demarcated from the body.

The species is extremely variable in colour (Fig. 39C, D, E). Individuals may be a more or less uniform brown or reddish-brown with a single pale yellow, cream or dirty white mid-dorsal stripe which may extend the full body length or be irregularly interrupted, but the marbled or banded varieties are more commonly encountered. In these the background colouration ranges from a pale yellowish-brown to cinnamon-brown or reddish-orange, speckled with brilliant white to yellowish flecks, and marked with transverse bands or irregular patches of dark brown, chestnut or brownish-yellow. There is sometimes a distinct brown lateral line on either side of the body. The ventral surface in all colour varieties is usually paler than the dorsal.

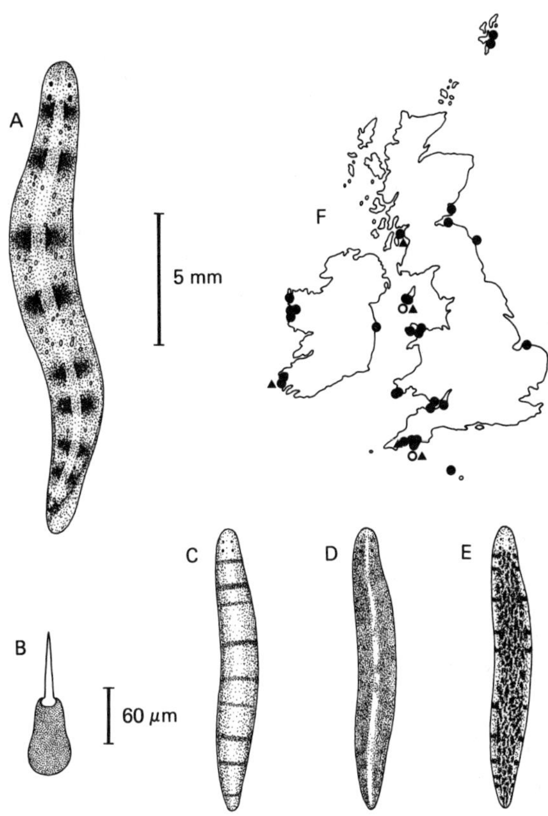

Fig. 39. *Oerstedia dorsalis*; A, dorsal view of whole animal, showing a typical colour pattern; B, a central stylet and basis; C–E, variations in dorsal colour pattern reported for this species; F, recorded distribution of *Oerstedia dorsalis* (●), *Oerstedia immutabilis* (○) and *Oerstedia nigra* (▲) from the British Isles. C redrawn from Bürger (1895).

Oerstedia dorsalis, although one of the smallest of the British nemerteans, is a common species with a distribution extending from the high neap tidal levels on the shore to 80 m or more sublittorally. It is found on a wide variety of substrata below low water mark (mud, gravel, sand, stones, shelly sediments) but generally occurs intertidally on small algae growing in rock pools (especially species of *Ceramium*, *Chondrus*, *Cladophora*, *Corallina* and *Ulva*) or between the holdfast branches of *Fucus* and *Laminaria* species. Occasional specimens have been found beneath intertidal stones, on *Zostera*, amongst ascidians or on the submerged surfaces of boats and hulks. A single record exists of an example found beneath a log in a poikilohaline lagoon.

Intertidal or shallow sublittoral *Oerstedia dorsalis* breed during the period September to November, but deeper-water forms have been found containing mature ova as early as June.

The species is widely distributed in the northern hemisphere and is found in the west Baltic Sea, the North Sea, the Mediterranean, on eastern Atlantic coasts from northern Europe to Madeira, and on both Atlantic and Pacific coasts of North America as far south as Mexico.

Oerstedia immutabilis (Riches, 1893)

Tetrastemma immutabile Riches, 1893; Herdman, 1894; Vanstone & Beaumont, 1894, 1895; Beaumont, 1895a, b; Sheldon, 1896

Oerstedia immutabilis M.B.A., 1904, 1931, 1957; Wijnhoff, 1912; Moore, 1937; Bruce *et al.*, 1963

Specific internal characters

Not known: although still listed as distinct species by Gibson (1982), Stiasny-Wijnhoff (1930) earlier concluded that this form, together with *Oerstedia nigra* (p. 148), could not with certainty be included in the genus *Oerstedia* because of the lack of data on their internal morphology. Both species have been established entirely on the basis of colour patterns which may well merely represent variations of the range of patterns reported for *Oerstedia dorsalis*, as suggested, for example, by Bürger (1895) and Joubin (1894).

Description

Inadequately described and in shape and size just like *Oerstedia dorsalis*, *Oerstedia immutabilis* is generally yellowish in colour but marked dorsally with chocolate-brown or orange-red pigment flecks which are strongly concentrated into a median dorsal line extending the full length of the body, i.e., a 'negative' version of the pattern illustrated for *Oerstedia dorsalis* in Fig. 39D. The four eyes are black.

This species is usually associated with *Oerstedia dorsalis* and is found intertidally and in shallow sublittoral locations among rock-pool corallines, laminarian holdfasts or on the fronds of smaller algae (see Fig. 39F).

Oerstedia nigra (Riches, 1893)

Tetrastemma nigrum Riches, 1893; Vanstone & Beaumont, 1894, 1895; Herdman, 1894, 1900; Beaumont, 1895a, b, 1900a, b; Sheldon, 1896
Oerstedia (*Tetrastemma*) *nigra* King, 1911
Oerstedia nigrum Southern, 1913
Oerstedia nigra M.B.A., 1904, 1931, 1957; Wijnhoff, 1912; Moore, 1937; Bruce *et al.*, 1963

Specific internal characters

Not known: see *Oerstedia immutabilis* (p. 147).

Description

5–15 mm long, with a general shape like *Oerstedia dorsalis*, *Oerstedia nigra* is ventrally a pale yellowish colour, dorsally dark brown to almost black due to a closely reticulated arrangement of pigment patches (somewhat resembling the *Oerstedia dorsalis* variety shown in Fig. 39E). A yellowish mid-dorsal stripe of very variable width may be present, or the entire dorsum may be uniformly coloured. The stripe, when present, is often irregularly interrupted by cross-bands of the darker pigment, giving an appearance close to that shown in Fig. 39D. The eyes of *Oerstedia nigra* are described as reddish in hue.

The ecological distribution of *Oerstedia nigra* parallels that of *Oerstedia dorsalis*, with which it is commonly found, the species generally being found among smaller algae or, less often, beneath stones or boulders from the lower shore to shallow sublittoral depths (Fig. 39F).

Genus *PROSORHOCHMUS* Keferstein, 1862

Diagnosis

Monostiliferous hoplonemerteans with shallow lateral or ventrolateral cephalic furrows; four eyes, the anterior pair usually larger than the posterior; rhynchocoel full body length, with wall composed of two distinct muscle layers; proboscis neither massive nor employed for locomotion; dermis moderately thick; body wall musculature thin, delicate; frontal organ a single exceptionally well developed tubular ciliated canal opening from tip of head; cephalic glands extensive, posteriorly reaching over cerebral ganglia, with at least two types of secretory cells; cerebral sensory organs small, anterior to cerebral ganglia; nervous system with neurochords and neurochord cells, without accessory lateral nerve; blood vascular system with three longitudinal vessels, without capillary network, with distinct valves, mid-dorsal vessel with one vascular plug; excretory system restricted to foregut region, with mononucleate flame cells which lack cuticular support bars; intestinal caecum present, with two anterior diverticula; hermaphroditic, occasionally or invariably ovoviviparous.

Prosorhochmus claparedii Keferstein, 1862
(Fig. 40)

? *Planaria flava* Montagu, 1808
Prosorhochmus claparedi M.B.A., 1957; Bruce *et al.*, 1963 (in part); Campbell, 1976
Prosorhochmus claparedei Pantin, 1961
Prosorhochmus claparedii Kerferstein, 1862; McIntosh, 1873–74, 1869; Sheldon, 1896; Pantin, 1969; Eason, 1973

Specific internal characters

Although several species of *Prosorhochmus* are reported (Gibson, 1982), *Prosorhochmus claparedii* is the only one adequately described; its specific internal characters are thus as given in the generic diagnosis.

Description

Prosorhochmus claparedii (Fig. 40A) is a small slender species, reaching a maximum length of about 35–40 mm. The head, sometimes appearing spatulate and marginally wider than the remaining body regions, frequently bears a distinct terminal notch which gives it a somewhat bilobed appearance. The four eyes are positioned just in front of the cerebral ganglia, the anterior pair being appreciably larger than the posterior pair. The body is rather flattened for most of its length and of a more or less uniform width until near the posterior tip, when it gradually tapers to end in a bluntly rounded tail.

The colour is typically a pale yellow or pale orange; the eyes are black and the position of the cerebral ganglia may be marked by a pair of greyish or translucent patches. The rhynchocoel is visible as a broad, pale mid-dorsal longitudinal band. In ripe individuals the young nemerteans can be seen through the parental body wall as whitish patches flanking the intestine.

Prosorhochmus claparedii, although predominantly an intertidal species dwelling beneath stones on coarse sediments or in rock crevices in the mid- to upper-shore regions, may be found by sublittoral dredging. In the Plymouth district it also extends to crevices at the top of the *Pelvetia* zone, where it is found along with collembolans, myriapods, chernetids and gastropods transitional to terrestrial forms (Pantin, 1969). It appears to be a gregarious species, since 'groups of adults (from 10–15 in number) are occasionally found in fissures' (McIntosh, 1873–74: 175). On Anglesey shores the species is most easily obtained during the reproductive period (June to August) and, when handled at this time, the adults readily give birth to several juvenile worms which emerge through the anus. McIntosh (1873–74) observed that the young may remain within the parent body until October.

The geographic distribution of the species extends from the Atlantic coast of France to the south and west coasts of England (Fig. 40C).

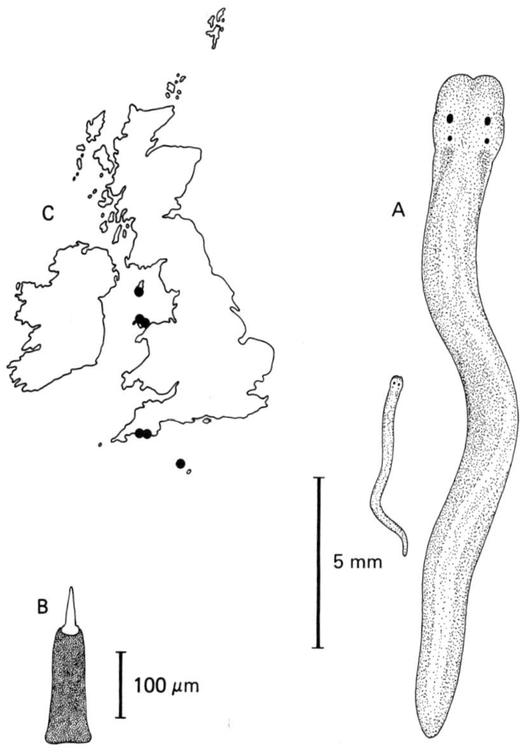

Fig. 40. *Prosorhochmus claparedii*; A, dorsal view of whole animal with a recently emerged juvenile alongside; B, a central stylet and basis; C, recorded distribution from the British Isles.

Family TETRASTEMMATIDAE

Genus *PROSTOMA* Dugès, 1828

Diagnosis

Small slender monostiliferous hoplonemerteans in which the body wall musculature is not strongly developed; eyes mostly 4–8, one species eyeless; rhynchocoel shorter than body in mature adults, with wall composed of two thin but distinct muscle layers; proboscis not strongly developed; frontal organ usually present, absent from one species; cephalic glands with variable degree of development depending upon the species; cerebral sensory organs small, anterior to cerebral ganglia; nervous system with neither neurochords nor neurochord cells, without accessory lateral nerve; blood vascular system with three longitudinal vessels, mid-dorsal vessel with one vascular plug; excretory system extending full length of body, with large numbers of nephridiopores; intestinal caecum either absent or very short, with anterior diverticula; hermaphroditic, true or protandrous; freshwater habits.

Prostoma graecense (Böhmig, 1892)
(Fig. 41A, B, C)

Tetrastemma graecensis Böhmig, 1892
Prostoma graecense Braithwaite & Clayton, 1945; Pantin, 1969; Gibson & Moore, 1976

Other reports of freshwater nemerteans from the British Isles, which may belong to this species but which cannot be positively identified (Gibson & Moore, 1976), have been recorded as:

Tetrastemma sp. Benham, 1892; Sheldon, 1896
Tetrastemma (*Emea*) sp. Mellanby, 1951
? *Tetrastemma aquarium dulcium* Sheldon, 1896
Prostoma sp. Stiasny-Wijnhoff, 1938
? *Prostoma clepsinoides* Southern, 1908b, 1911

Specific internal characters

Oesophagus distinct and ciliated; cephalic glands extending back to cerebral ganglia; rhynchodaeum with well developed longitudinal muscles; proboscis with 9–10 nerves.

Description

Prostoma graecense (Fig. 41A, C) is a small and slender nemertean, rarely exceeding 20 mm in length, characteristically found crawling on aquatic plants or on the surface of mud. Adults are typically coloured a uniform reddish-brown, although bright orange, brown, red or greenish individuals have been reported. Juvenile worms are commonly white to pale yellowish or

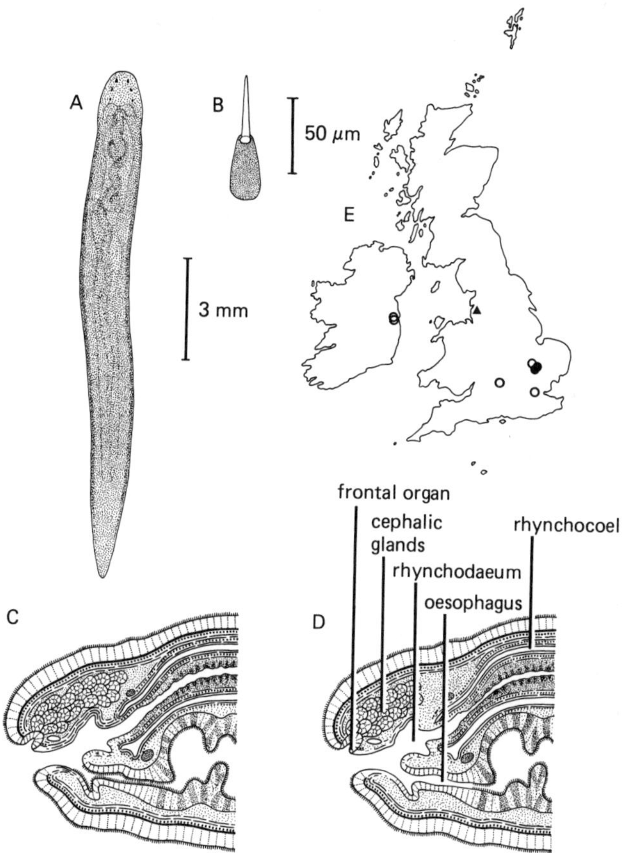

Fig. 41. *Prostoma graecense*, *Prostoma jenningsi*; A, dorsal view of whole animal (*note*; both species are identical in their external appearance); B, a central stylet and basis; C, D, schematic vertical longitudinal sections through the anterior regions of *Prostoma graecense*, C, and *Prostoma jenningsi*, D, to illustrate structures of taxonomic importance for distinguishing between these freshwater nemertean species; E, recorded distribution of *Prostoma graecense* (●), *Prostoma jenningsi* (▲) and *Prostoma* sp. (○) from the British Isles. C, D redrawn from Gibson & Moore (1976).

straw coloured. Eye number varies between four and six depending upon age and size.

The species possesses a world-wide though sporadic distribution.

Prostoma graecense and the other British member of this genus, *Prostoma jenningsi* (p. 154), cannot be distinguished apart on their external features alone and must be investigated histologically.

Prostoma jenningsi Gibson & Young, 1971
(Fig. 41A, B, D)

Prostoma jenningsi Gibson & Young, 1971, 1976; Gibson & Moore, 1976

Specific internal characters

Oesophagus distinct but not ciliated; cephalic glands not posteriorly reaching cerebral ganglia; rhynchodaeum with weakly developed longitudinal musculature; proboscis with 11 nerves.

Description

In size, shape and general external appearance *Prostoma jenningsi* (Fig. 41A, D) closely resembles *Prostoma graecense* and can only be distinguished from it by its internal morphology.

Prostoma jenningsi is thus far known from only a single locality (Fig. 41E).

Genus *TETRASTEMMA* Ehrenberg, 1831

Diagnosis

Mostly small slender monostiliferous hoplonemerteans in which the body wall musculature is not strongly developed; eyes usually distinct, mostly four arranged at the corners of a square or rectangle, occasionally fragmented; rhynchocoel extending to or almost to the posterior end of the body, with wall containing two distinct muscle layers; proboscis not strongly developed; frontal organ probably present; cephalic glands usually well developed but rarely extending behind cerebral ganglia; cerebral sensory organs small, close to anterior margins of cerebral ganglia; nervous system with neither neurochords nor neurochord cells, without accessory lateral nerve; blood vascular system with three longitudinal vessels, mid-dorsal vessel with one (?) vascular plug; excretory system restricted to foregut regions, with two or only a few nephridiopores; intestinal caecum present, with anterior diverticula; sexes separate; marine or estuarine.

Many of the 80 or more described species of *Tetrastemma* (Gibson, 1982) are distinguished only by their colour patterns and little data are available on their internal anatomy. It is more than probable that several species belong to other genera and/or represent colour varieties of previously established forms. Even where the internal morphology has been investigated (Kirsteuer, 1963b), the taxonomic significance of such features as the organisation of the stomach and pyloric duct, the origin of the mid-dorsal blood vessel and the relationships between other structures remain uncertain and often appear to be highly variable at the intraspecific level.

Tetrastemma ambiguum Riches, 1893
(Fig. 42A, B)

Tetrastemma ambiguum Riches, 1893; Sheldon, 1896; M.B.A., 1904, 1957
Prostoma ambiguum Wijnhoff, 1912; M.B.A., 1931

Specific internal characters

Not known; this species is inadequately described.

Description

A slender-bodied species 10–15 mm long, *Tetrastemma ambiguum* (Fig. 42A) possesses a distinct head which is broader than the remaining body regions and bears two pairs of cephalic furrows. The four eyes are typically brown in colour, irregularly shaped and not well defined, although in one example they are recorded as distinct and black. The anterior pair of eyes is at least twice the size of the posterior pair.

In colour *Tetrastemma ambiguum* is an overall pale yellow, but is usually marked on the dorsal surface with a variable amount of reddish-brown pigmentation.

Known only from the British Isles (Fig. 45A), the species is found sublittorally at depths to about 60 m on a variety of sediment types (sand, mud, limestone fragments, stones). Sexually mature individuals have been found in November.

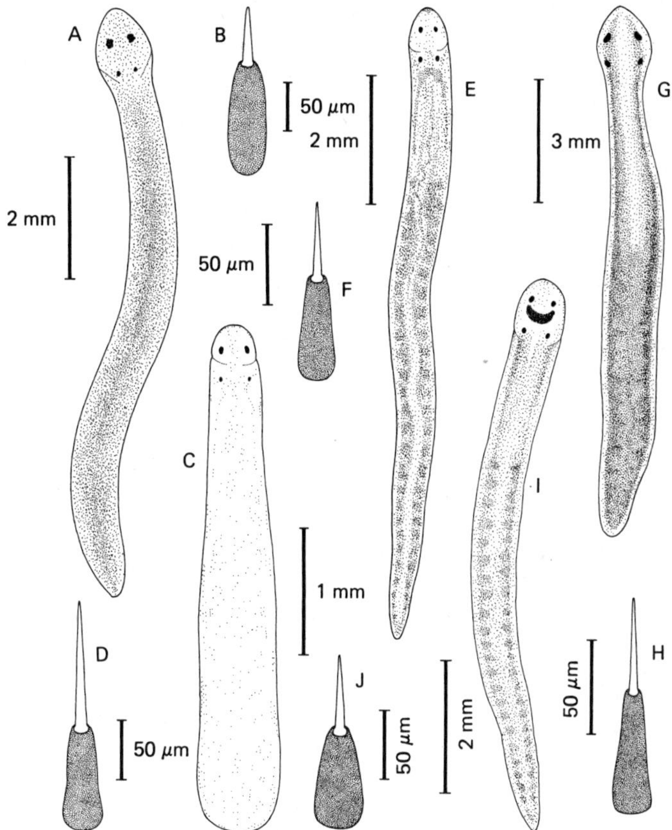

Fig. 42. *Tetrastemma ambiguum*; A, dorsal view of whole animal; B, a central stylet and basis. *Tetrastemma beaumonti*; C, dorsal view of whole animal; D, a central stylet and basis. *Tetrastemma candidum*; E, dorsal view of whole animal; F, a central stylet and basis. *Tetrastemma cephalophorum*; G, dorsal view of whole animal; H, a central stylet and basis. *Tetrastemma coronatum*; I, dorsal view of whole animal; J, a central stylet and basis. (*Note*: amongst the tetrastemmids there is often a considerable variation in the form of the stylet armature; those illustrated are merely examples for the appropriate species). A, B based on description given by Riches (1893), C, D redrawn from Southern (1913), G–J redrawn from Bürger (1895).

Tetrastemma beaumonti (Southern, 1913)
(Fig. 42C, D)

Prostoma beaumonti Southern, 1913; Farran, 1915

The generic names *Prostoma* and *Tetrastemma* were commonly regarded as synonymous until Stiasny-Wijnhoff (1938) discriminated between them by uniting all the freshwater species into the genus *Prostoma*. Since that time, marine or estuarine '*Prostoma*' have been included in the genus *Tetrastemma*; *Prostoma beaumonti* Southern, 1913, is accordingly listed here as *Tetrastemma beaumonti* (Southern, 1913).

Specific internal characters

Not known; the species is inadequately described.

Description

Tetrastemma beaumonti (Fig. 42C) is a small form, not exceeding 5–6 mm in extended length and usually only 3 mm long. The body is stout, cylindrical and thickest at its bluntly rounded posterior end. It gently tapers towards the head, which is truncated, narrower than the body and posteriorly marked by shallow transverse cephalic furrows. The four distinct eyes are arranged at the corners of a rectangle, the anterior pair usually, though not invariably, being the larger.

In colour *Tetrastemma beaumonti* is a creamish-white, occasionally tinged pink, without a distinct colour pattern. Under a microscope irregularly distributed and shining epidermal glands are usually evident in life.

Sexually mature specimens, containing 6–8 pairs of gonads distributed between the intestinal diverticula, are found during the period March to September. Ripe gonads are present in individuals as small as 2 mm in length.

This is a sublittoral species, dredged from 3–20 m depth on gravelly and sandy sediments. It has only been recorded from the Atlantic coast of Ireland (Fig. 45A), although Southern (1913) suggests that Beaumont's (1895a, b) record of *Tetrastemma candidum* from the Isle of Man may be the same species.

Tetrastemma candidum (Müller, 1774)
(Fig. 42E, F)

Fasciola candida Müller, 1774
Planaria quadrioculata Johnston, 1828–29
Nemertes Nemertes quadrioculata Johnston, 1837
Prostoma quadrioculata Johnston, 1846
Planaria algae Dalyell, 1853
? *Vermiculus coluber* var. Dalyell, 1853
Polia quadrioculata Williams, 1858; McIntosh, 1869
Tetrastemma varicolor Claparède, 1862; Johnston, 1865 (in part); Lankester, 1866; Parfitt, 1867; McIntosh, 1869
Tetrastemma? algae Johnston, 1865
Tetrastemma algae McIntosh, 1869
Tetrastemma candida McIntosh, 1873–74, 1875c; Haddon, 1886a
Prostoma (Tetrastemma) candidum Southern, 1908a; King, 1911; Stephenson, 1911
Prostoma candidum Wijnhoff, 1912; Southern, 1913; Farran, 1915; M.B.A., 1931; Moore, 1937
Tetrastemma candidum Koehler, 1885; Riches, 1893 (in part); Vanstone & Beaumont, 1894, 1895; Herdman, 1894, 1900; Beaumont, 1895a, b (in part), 1900a, b; Gamble, 1896; Sheldon, 1896; Jameson, 1898; Allen, 1899; M.B.A., 1904, 1957; Horsman, 1938; Bassindale & Barrett, 1957; Bruce *et al.*, 1963; Crothers, 1966; Laverack & Blackler, 1974

Specific internal characters

Uncertain; according to Kirsteuer (1963b) diagnostic features are the mid-dorsal blood vessel arising from a transverse connective linking the two cephalic vessels, and the absence of distinct in- or outfoldings from the stomach, which leads directly into the pyloric duct.

Description

Tetrastemma candidum (Fig. 42E) lacks distinct external characters and, indeed, is an inadequately described species; many of the records of the species must thus be of uncertain validity. It is possible that several species have been listed under the same name. Among the British tetrastemmids it may in particular be confused with *Tetrastemma flavidum* (p. 163), although unlike this form *Tetrastemma candidum* is a very restless and actively moving species.

Descriptions of *Tetrastemma candidum* are far from being in agreement. Mostly reported as about 8–10 mm long and 1 mm or less wide, individuals with lengths in the 20–35 mm range have been ascribed to this species. The head is rather flattened, bluntly rounded and may be wider or narrower than the succeeding body regions. It bears four distinct reddish-brown, dark brown or black eyes. Posteriorly the body gradually narrows and ends in a blunt point.

The colour is apparently extremely variable, examples being recorded as pale dull yellow, deep apricot yellow, dull brownish-orange, light orange, pale greyish-yellow, light brownish-grey, brown, deep orange-red, reddish-brown, pale green or greenish-yellow. Much of this variation appears to be attributable to gut contents, as pale flesh-coloured or yellowish individuals have been described in which only the intestine is distinctly pink, brick-red or bright green. Colour variations may also be due to different species being recorded as *Tetrastemma candidum*. Immature specimens, without obvious gut contents, possess a characteristic greyish-white translucent appearance with a faint greenish tinge and Coe (1943) regards a pale green colour as typical for the species. Ripe gonads may appear as a grey or yellow tinge. On the head an opaque whitish patch is sometimes present between the two pairs of eyes, and from this a slender median white longitudinal streak occasionally extends down the back. The cephalic furrows may be colourless or possess a brown pigmentation.

Occurring from about mid-shore level down to depths of 55 m or more, *Tetrastemma candidum* is commonly obtained from rock pool algae (especially *Ulva*, *Corallina* and *Cladophora*), from within the bladders of the wrack *Ascophyllum nodosum* (Linnaeus) Le Jolis, with *Zostera*, among colonies of hydroid coelenterates, in old tubes of sabellid polychaetes, on submerged surfaces of hulks and boats, under rocks or in sediments such as shelly gravel and sand. The gonads are ripe in late spring and summer (April to August).

Tetrastemma candidum has a circumpolar distribution in the northern hemisphere. Apart from the British Isles (Fig. 45B), it is recorded from the coasts of Scandinavia, the North Sea, the Mediterranean, Madeira, the Faroes, Iceland, Greenland, the Atlantic and Pacific coasts of North America, the Caribbean and Japan. In the southern hemisphere examples identified as members of this species are reported from South Africa and Brazil.

Tetrastemma cephalophorum Bürger, 1895
(Fig. 42G, H)

Prosorhochmus claparedii Riches, 1893; Beaumont, 1895a, b; Herdman, 1896, 1900
Tetrastemma cephalophorum Bürger, 1895; Beaumont, 1900a, b; M.B.A., 1904, 1957
Prostoma cephalophorum Wijnhoff, 1912; Southern, 1913; M.B.A., 1931
Prosorhochmus claparedi Moore, 1937; Bruce *et al.*, 1963 (in part)

Specific internal characters

Not known; this species is inadequately described.

Description

Tetrastemma cephalophorum (Fig. 42G) is up to about 15 mm long and 1.5–2.0 mm wide, with a distinct and rather diamond-shaped head, wider than the body, bearing four very large eyes. The central stylet of the proboscis is very long and slender, the stylet basis posteriorly truncate and narrow (Fig. 42H).

In colour *Tetrastemma cephalophorum* is dorsally a uniform reddish-brown, but the head, ventral surface and lateral body margins are a pale yellowish hue.

Typically dredged from coarse shell gravel or stony sediments at depths of 10–15 m, the geographic distribution of the species extends from the British Isles (Fig. 45C) to the Mediterranean.

Tetrastemma coronatum (Quatrefages, 1846)
(Fig. 42I, J)

Polia coronata Quatrefages, 1846
Tetrastemma melanocephalum Riches, 1893 (in part); Beaumont, 1895a, b, 1900a, b (in part); M.B.A., 1904 (in part)
Tetrastemma melanocephalum var. *diadema* Gamble, 1896
Prostoma coronatum Wijnhoff, 1912; Southern, 1913 (in part); Farran, 1915; M.B.A., 1931
Tetrastemma coronatum M.B.A., 1957; Green, 1968

Specific internal characters

Several authors have regarded *Tetrastemma coronatum*, together with *Tetrastemma diadema* Hubrecht, 1879, not certainly recorded from British waters, merely as a colour variety of the related species, *Tetrastemma melanocephalum* (p. 168). Kirsteuer (1963b), however, regards these as three distinct species which can be separated by internal features as well as by their colour patterns. *Tetrastemma coronatum* is characterised by its mid-dorsal blood vessel arising from a branch of one lateral vessel and by its complex foregut organisation in which the stomach is more or less subdivided into two regions by a distinct fold, the posterior portion leading directly into the pyloric tube.

Description

Typically up to some 12–15 mm in length and 0.5–1.0 mm in width, *Tetrastemma coronatum* (Fig. 42I) possesses a blunt head which is not distinct from the body although its posterior region bears a pair of shallow transverse cephalic furrows. The four eyes are moderately sized and obvious.

The general colour is pale yellowish-green, light green or light brownish-green; Bürger (1895) observed that mature females were always larger than males and greenish coloured, whereas the males exhibited a brownish tinge. In both sexes the head is almost colourless apart from the dorsal dark-brown or blackish crescent-shaped pigment patch characteristic of the species. Just in front of and behind this crescent there may be whitish or opaque regions, occasionally extending posteriorly as a slender mid-dorsal streak.

Tetrastemma coronatum is found amongst intertidal algae, often in the holdfasts of *Laminaria*, with *Zostera*, on bryozoans, with tubicolous polychaetes, in sandy detritus or on stones dredged to depths of about 40 m. Sexually mature examples may be found during July and August.

The geographic range of this species extends from the British Isles (Fig. 45D) to Scandinavia, the Atlantic coast of France, Madeira, the Mediterranean and the Black Sea.

Tetrastemma flavidum Ehrenberg, 1831
(Fig. 43A, B)

Tetrastemma flavidum Ehrenberg, 1831; Beaumont, 1895a, b?; Sheldon, 1896; Herdman, 1896, 1900?; Jameson, 1898; Allen, 1899; M.B.A., 1957; Bruce *et al.*, 1963?; Crothers, 1966; Williams, 1972
Polia sanguirubra McIntosh, 1869; Koehler, 1885
Tetrastemma flavida Haddon, 1886a; Elmhirst, 1922; Eales, 1952
Tetrastemma candidum Riches, 1893 (in part)
Prostoma flavidum Southern, 1908a, 1913; Wijnhoff, 1912; Farran, 1915; M.B.A., 1931; Moore, 1937?; Barrett & Yonge, 1958

Records of this species from the British Isles are confused. *Tetrastemma flavida* as reported by McIntosh (1873–74) is not the same species as that originally described by Ehrenberg (1831), but refers to another monostiliferous hoplonemertean now known as *Nemertopsis flavida* (p. 138). Accordingly authors' records relying upon McIntosh's work as the naming authority are listed under *Nemertopsis*, whereas those either using Bürger (1895) as the diagnostic source or not specifying an authority are itemised above; Wijnhoff (1912) clearly indicates that *Tetrastemma flavidum sensu* Bürger (1895) is the same as Ehrenberg's (1831) species but differs from *Tetrastemma flavidum sensu* McIntosh (1873–74).

Specific internal characters

Not a well characterised species, Kirsteuer (1963b) lists as specific features the way in which the mid-dorsal blood vessel arises from a transverse connective linking the two cephalic vessels, and the manner whereby the dorsal wall of the stomach is distinctly infolded and the stomach leads directly into the pyloric tube.

Description

Tetrastemma flavidum (Fig. 43A) reaches lengths of 14–15 mm but is generally only 0.5–0.75 mm wide. The body is rather flattened, bluntly rounded at its posterior tip and either rounded or slightly tapered at the head. The four eyes are very small but usually distinct. In colour typically a bright pink, smaller individuals may be tinged pale yellowish or reddish and appear more or less transparent with their various internal organs clearly visible in transmitted light. The lateral margins are translucent.

The species is found in mud, fine sand or gravel, or among the branches of laminarian holdfasts, from the lower shore to depths of up to 100 m. Examples found infrequently in the branchial cavity of ascidians such as *Ascidia mentula* Müller may represent a different species or merely be in the 'host' by chance; there is no evidence at present to suggest that the species possesses even occasional commensal or parasitic habits. Mature individuals are found during the summer months.

The geographic range extends from the British Isles (Fig. 46A) and Scandinavia to the Mediterranean and Red Sea.

Tetrastemma helvolum Bürger, 1895
(Fig. 43C)

Tetrastemma candidum Riches, 1893 (in part); Beaumont, 1895a, b (in part)
Tetrastemma helvolum Bürger, 1895; M.B.A., 1957
Prostoma coronatum var. Southern, 1913 (in part)
Prostoma helvolum Wijnhoff, 1912; M.B.A., 1931

Specific internal characters

Tetrastemma helvolum, which is an inadequately described form, is regarded by some authors as a colour variety of *Tetrastemma candidum* rather than as a distinct species. Kirsteuer (1963b), however, regards as specific the way in which the mid-dorsal blood vessel arises from one lateral vessel and by the manner in which the pyloric tube emerges from a dorsomedian outfold at the back of the stomach.

Description

Tetrastemma helvolum (Fig. 43C) is up to 20 mm long but less than 1 mm wide. The body is bluntly pointed at its posterior end and a bright pale honey-yellow colour apart from the paler yellowish head. Characteristically at the tip of the head, between the anterior pair of eyes, and in the anal region, dense accumulations of gland cells appear as conspicuous shining whitish masses.

The species, which is recorded from the British Isles (Fig. 46B) and the Mediterranean, can be dredged from 4–80 m depth from coralline, muddy, shelly or sandy sediments or amongst algae.

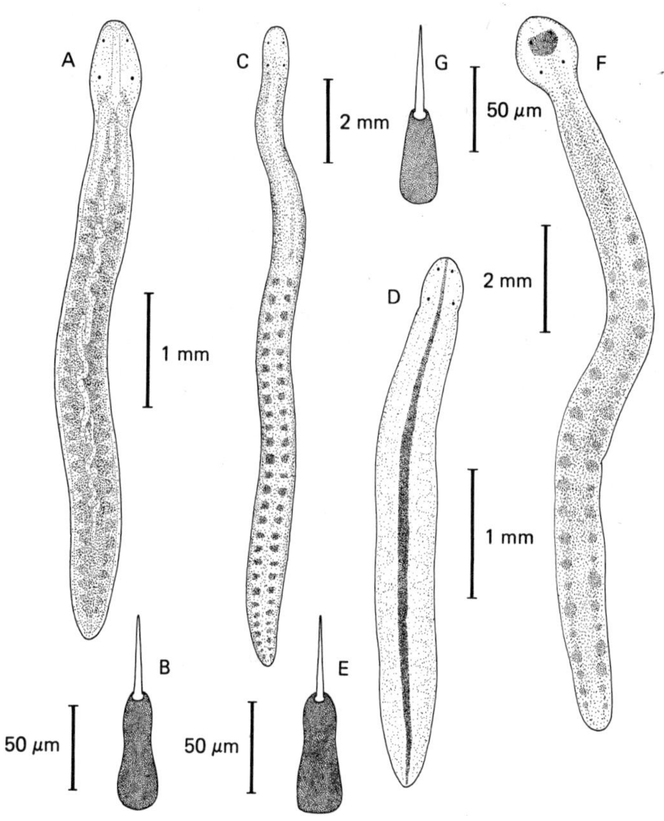

Fig. 43. *Tetrastemma flavidum*; A, dorsal view of whole animal; B, a central stylet and basis. *Tetrastemma helvolum*; C, dorsal view of whole animal. *Tetrastemma herouardi*; D, dorsal view of whole animal; E, a central stylet and basis. *Tetrastemma longissimum*; F, dorsal view of whole animal; G, a central stylet and basis. A–C, F, G redrawn from Bürger (1895), D, E redrawn from Oxner (1908).

Tetrastemma herouardi (Oxner, 1908)
(Fig. 43D, E)

Prostoma herouardi Oxner, 1908; Wijnhoff, 1912; M.B.A., 1931
Tetrastemma herouardi M.B.A., 1957

Specific internal characters

Cerebral sensory organs as large as and linked by distinct nerves to dorsal cerebral ganglionic lobes; cephalic glands well developed.

Description

A small species, up to 5–6 mm long and 0.5–0.75 mm wide, *Tetrastemma herouardi* (Fig. 43D) is a transparent light flesh colour marked with a single mid-dorsal longitudinal stripe of a dark wine-red hue extending the full length of the body.

In the British Isles (Fig. 46C) it has been found on only one occasion on a ship hulk, but at Roscoff, France, it occurs on brown algae of the genus *Cystoseira* and with the tunicate *Dendrodoa grossularia* (Van Beneden).

Tetrastemma longissimum Bürger, 1895
(Fig. 43F, G)

Tetrastemma longissimum Bürger, 1895; M.B.A., 1957
? *Tetrastemma flavidum* var. *longissimum* Gamble, 1896
Prostoma longissimum Wijnhoff, 1912; M.B.A., 1931

Specific internal characters

An inadequately described form, this species somewhat resembles *Tetrastemma coronatum* (p. 162) but differs in both the colour and shape of the cephalic pigment spot. The origin of the mid-dorsal blood vessel is the same in both species, but whereas *Tetrastemma coronatum* has a complex foregut arrangement, in *Tetrastemma longissimum* the stomach is simple and leads directly to the pyloric duct (Kirsteuer, 1963b).

Description

Although *Tetrastemma longissimum* (Fig. 43F) may attain a length of more than 20 mm, it has a slender body, posteriorly rounded, of about 1 mm width. The broad, rounded head is generally distinct from the body; it is almost colourless apart from a characteristic transverse band of bright red, dark red or shining reddish-brown pigment positioned between the four distinct eyes. The remaining body colour is brownish-yellow, although in sexually mature individuals (July and August) the green tinged gonads are clearly evident.

The species occurs intertidally at Naples, but in the Adriatic and around the British Isles (Fig. 46C) has only been found sublittorally at depths to about 20 m. It occurs among smaller algae or on shell and sand detritus sediments.

Tetrastemma melanocephalum (Johnston, 1837)
(Fig. 44A, B)

? *Planaria ascaridea* Montagu, 1808
Planaria unipunctata Montagu, 1808
Nemertes Nemertes melanocephala Johnston, 1837
Nemertes melanocephala Thompson, 1846
Prostoma melanocephala Johnston, 1846; Thompson, 1856
Vermiculus coluber Dalyell, 1853
Omatoplea melanocephala Johnston, 1865; Parfitt, 1867
Ommatoplea melanocephala Lankester, 1866; McIntosh, 1868a, 1869
Cephalotrix unipunctata Parfitt, 1867
Tetrastemma melanocephala McIntosh, 1873–74, 1875c
Tetrastemma melanocephalum Riches, 1893 (in part); Vanstone & Beaumont, 1894, 1895; Herdman, 1894, 1900; Beaumont, 1895a, b, 1900a, b (in part); Sheldon, 1896; Jameson, 1898; M.B.A., 1904, 1957; Horsman, 1938; Bassindale & Barrett, 1957; Bruce *et al.*, 1963; Crothers, 1966; Gibson & Jennings, 1967; Gibson, 1968a; Jennings & Gibson, 1969; Laverack & Blackler, 1974; Campbell, 1976; Boyden *et al.*, 1977
Prostoma melanocephalum Wijnhoff, 1912; Southern, 1913; Farran, 1915; M.B.A., 1931; Moore, 1937

Specific internal characters

Mid-dorsal blood vessel arising from left lateral vessel; stomach with distinct dorsal infolding and leading directly into pyloric duct (Kirsteuer, 1963b).

Description

Tetrastemma melanocephalum (Fig. 44A) is probably the largest of the British tetrastemmids, attaining a length of 30–60 mm and width of 2.0–2.5 mm when sexually mature. The general colour is typically yellow or yellowish-green, sometimes more reddish-brown, with a characteristic dark brown or black quadrangular pigment patch on the head. Occasional individuals may be found with minute brownish pigment flecks distributed along the lateral body margins.

Common but rarely abundant (R. S. K. Barnes, personal communication reports that up to 500/m^2 individuals can be found in sandy beds of salt-marsh creeks, preying on the amphipod *Corophium arenarium*, on and near Scolt Head Island, Norfolk), *Tetrastemma melanocephalum* is most easily obtained by standing rock pool algae (especially species of *Ceramium*, *Corallina*, *Cladophora* and *Fucus vesiculosus*) in sea water without aeration; the nemerteans usually emerge within 24–48 hours to creep around the water surface. The species is also found intertidally in wet sand, in rock crevices, between laminarian holdfasts and on other fucoid algae. Sublittorally it occurs at depths to about 40 m in a wide range of habitats, including submerged ship hulks, mud, shell fragments, stones, gravel and *Zostera*. It is

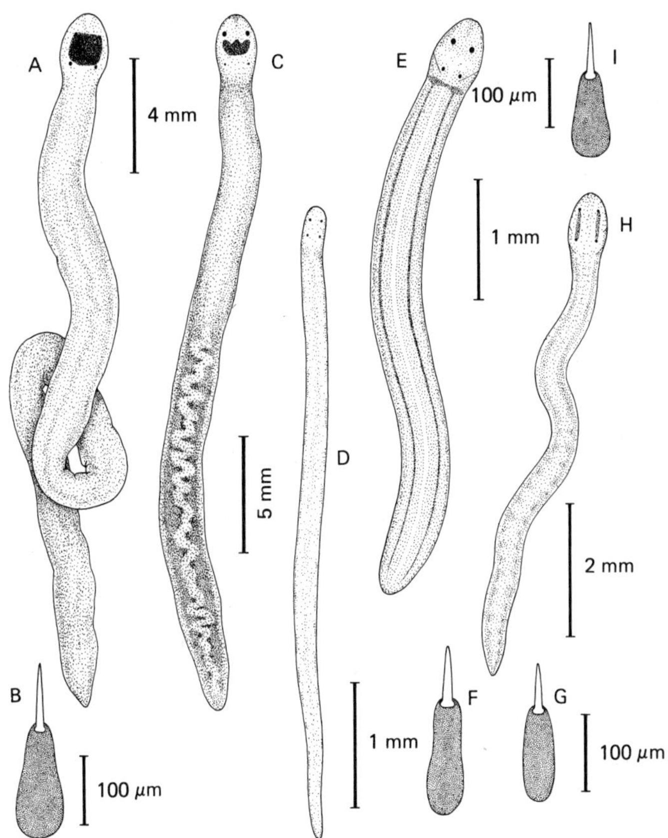

Fig. 44. *Tetrastemma melanocephalum*; A, dorsal view of whole animal; B, a central stylet and basis. *Tetrastemma peltatum*; C, dorsal view of whole animal. *Tetrastemma quatrefagesi*; D, dorsal view of whole animal. *Tetrastemma robertianae*; E, dorsal view of whole animal; F, G, two forms of central stylet and basis. *Tetrastemma vermiculus*; H, dorsal view of whole animal; I, a central stylet and basis. C, H redrawn from Bürger (1895), D based on description given by Bürger (1904), E–G redrawn from Berg (1973).

sometimes found under conditions of reduced salinity, as occur in the lower reaches of estuaries or in Kiel Bay, Germany (Brunberg, 1964).

The geographic range of *Tetrastemma melanocephalum* extends from the British Isles (Fig. 46D) to Scandinavia, Madeira, the Canary Islands, the Mediterranean and the coasts of the Black Sea.

Tetrastemma peltatum Bürger, 1895
(Fig. 44C)

Tetrastemma peltatum Bürger, 1895; M.B.A., 1957
Prostoma peltatum Wijnhoff, 1912; M.B.A., 1931

Specific internal characters

An insufficiently well characterised species, Bürger (1895) separated *Tetrastemma peltatum* from *Tetrastemma coronatum*, which resemble each other, through its much larger size and by the massive development of the cephalic glands. Kirsteuer (1963b) records the simple stomach leading directly into the pyloric duct and the mid-dorsal blood vessel emerging from a transverse connective, and not forming a vascular plug, as specific characters.

Description

Tetrastemma peltatum (Fig. 44C) is a fairly bulky tetrastemmid, up to 40–50 mm long and almost 2 mm wide. The body is not posteriorly tapered and ends in a bluntly rounded tail. The head is oval in shape and slightly wider than the adjacent body regions, with the eyes arranged into large anterior and small posterior pairs; Wijnhoff (1912) regards the unequal size of the eyes as a diagnostic character. Fawn to deep brown in overall colour, the intestine commonly appears greenish, presumably due to gut contents. On the head a large broad crescentic patch of brown or black pigment lies about mid-way between the two pairs of eyes; the shape of this patch differs from that of *Tetrastemma coronatum* in having a median anterior point projecting from the arc of the crescent.

Only two specimens of *Tetrastemma peltatum* have been found in British waters (Fig. 46C). The species is otherwise known mainly from the Mediterranean and Adriatic, although it is also recorded from the coast of Chile.

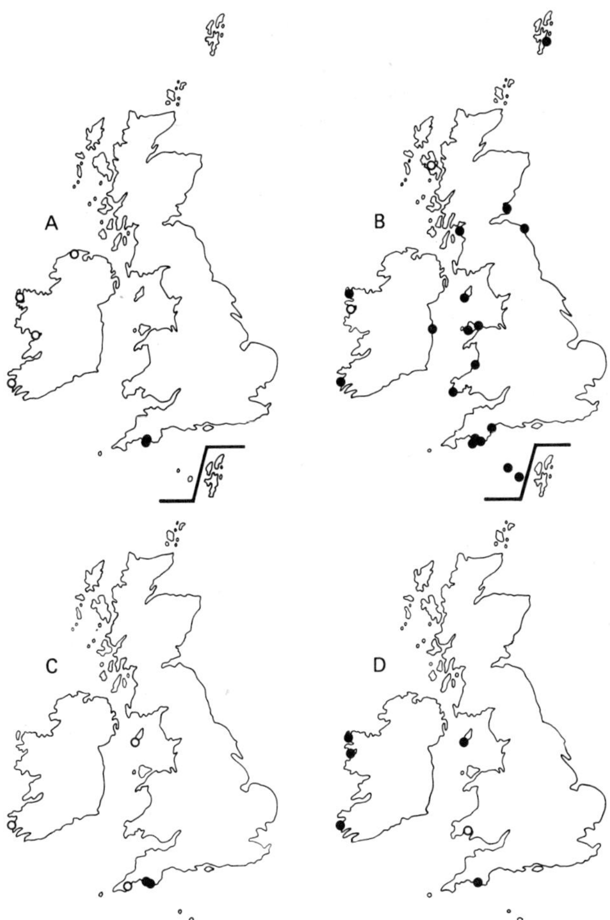

Fig. 45. A, recorded distribution of *Tetrastemma ambiguum* (●) and *Tetrastemma beaumonti* (○) from the British Isles; B, recorded distribution of *Tetrastemma candidum* from the British Isles; ● = reasonably certain records, ○ = specific identity or location uncertain; C, recorded distribution of *Tetrastemma cephalophorum* from the British Isles; ● = certain records, ○ = specific identity uncertain; D, recorded distribution of *Tetrastemma coronatum* from the British Isles; ● = certain records, ○ = specific identity uncertain.

Tetrastemma quatrefagesi Bürger, 1904
(Fig. 44D)

Tetrastemma quatrefagesi Bürger, 1904; M.B.A., 1957
Prostoma quatrefagesi Wijnhoff, 1912; M.B.A., 1931

Specific internal characters

Not known; this species is very poorly described.

Description

Tetrastemma quatrefagesi (Fig. 44D) is 5–6 mm long with a cylindrical body. The colour is a transparent yellow without external markings, the internal organs being easily visible under transmitted light. The presence of four accessory stylet pouches in the proboscis appears to be the only feature distinguishing this form from *Tetrastemma flavidum* (p. 163); whether or not this character is taxonomically significant is at present unknown.

Only a single specimen of this species has been recorded from the British Isles (Fig. 46C), sublittorally from shelly gravel. The geographic range extends to the Mediterranean.

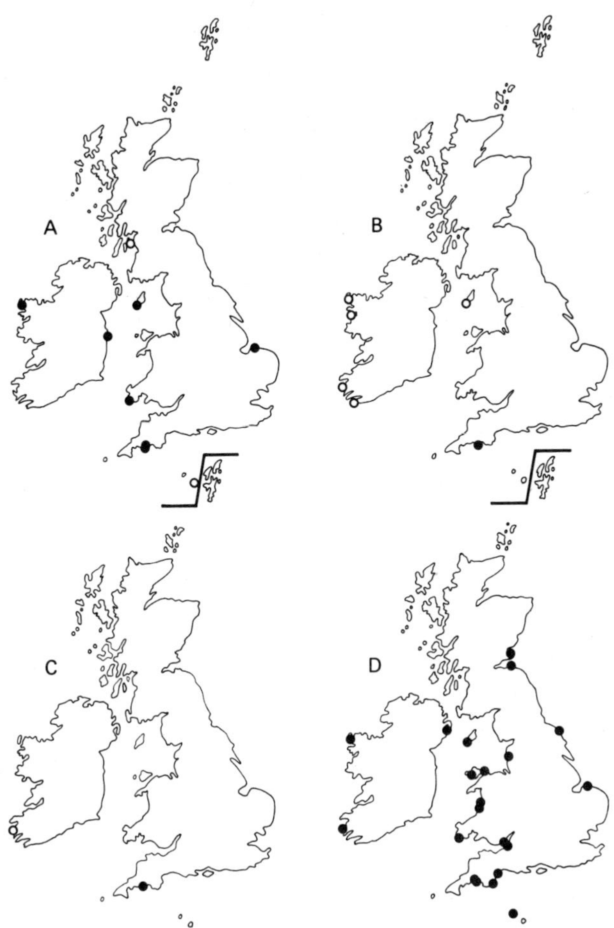

Fig. 46. A, recorded distribution of *Tetrastemma flavidum* from the British Isles; ● = reasonably certain records, ○ = specific identity uncertain; B, recorded distribution of *Tetrastemma helvolum* from the British Isles; ● = reasonably certain record, ○ = specific identity uncertain; C, recorded distribution of *Tetrastemma herouardi*, *Tetrastemma longissimum*, *Tetrastemma peltatum* and *Tetrastemma quatrefagesi* (●) from the British Isles; ○ = uncertain record of *Tetrastemma longissimum*; D, recorded distribution of *Tetrastemma melanocephalum* from the British Isles.

Tetrastemma robertianae McIntosh, 1873–74
(Fig. 44E, F, G)

Tetrastemma robertianae McIntosh, 1873–74; Riches, 1893; Vanstone & Beaumont, 1894, 1895; Herdman, 1894, 1900; Beaumont, 1895a, b, 1900a, b; Sheldon, 1896; M.B.A., 1957; Bruce *et al.*, 1963

Prostoma robertianae King, 1911; Wijnhoff, 1912; Southern, 1913; M.B.A., 1931; Moore, 1937

Specific internal characters

Cephalic glands well developed, posteriorly extending to cerebral ganglia; dorsal cerebral ganglionic lobes small; cerebral sensory organs moderately large.

Description

Tetrastemma robertianae (Fig. 44E), which is up to 30–35 mm long and 0.7–1.0 mm wide, is a strikingly and characteristically coloured species fully described recently by Berg (1973). The background colour is orange, pinkish-brown or yellowish, marked with a transverse brown band which encircles the body at the rear of the head. This band may be ventrally incomplete. From the rear of the band a pair of dorsolateral brown stripes run toward and may meet at the tail. A mid-dorsal longitudinal stripe of white pigment extends posteriorly between the brown stripes, and white pigment patches also occur on the head between the two pairs of eyes; sometimes this white cephalic pigment is developed into a broad transverse band. The eyes are black, the anterior pair being significantly larger than the posterior. The ventral surface is a pale pinkish brown. The body is rather flattened and bluntly rounded at its posterior tip. The head is rhomboidal in shape, wider than the succeeding body regions and anteriorly blunt.

A sublittoral species, *Tetrastemma robertianae* is found at depths to 70 m or more on mud, shelly gravel and stones, or in shallow waters among laminarian holdfasts. The geographic range extends from Scandinavia to the British Isles (Fig. 47A).

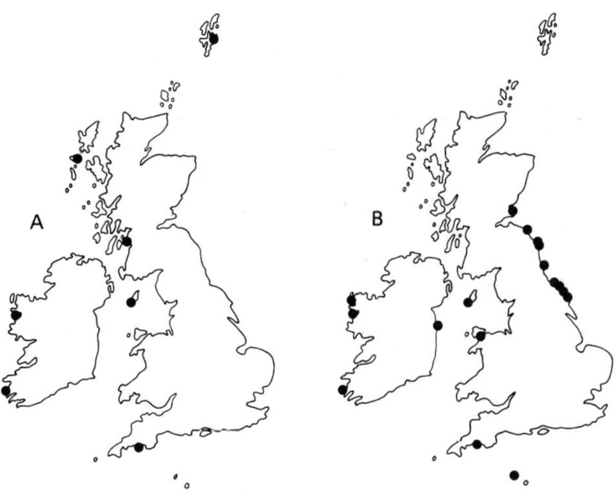

Fig. 47. A, recorded distribution of *Tetrastemma robertianae* from the British Isles; B, recorded distribution of *Tetrastemma vermiculus* from the British Isles.

Tetrastemma vermiculus (Quatrefages, 1846)
(Fig. 44H, I)

Polia vermiculus Quatrefages, 1846

Tetrastemma vermiculus McIntosh, 1869; Jameson, 1898; M.B.A., 1957; Bruce *et al.*, 1963; Laverack & Blackler, 1974; Knight-Jones & Nelson-Smith, 1977

Tetrastemma vermicula McIntosh, 1873–74, 1875c

Tetrastemma vermiculatum Riches, 1893; Herdman, 1894; Vanstone & Beaumont, 1894, 1895; Beaumont, 1895a, b, 1900a, b; Gamble, 1896; Sheldon, 1896; M.B.A., 1904

Prostoma vermiculus Southern, 1908a, 1913; Wijnhoff, 1912; Farran, 1915; M.B.A., 1931; Moore, 1937; Bruce, 1948

Prostomatella vermicula Moore, 1973

Specific internal characters

This is an inadequately described species which was transferred to the genus *Prostomatella* by Friedrich (1935), but subsequently returned by him, with some reservations, to *Tetrastemma* (Friedrich, 1955). Kirsteuer (1963b) records as specific features a mid-dorsal blood vessel, which arises from a transverse connective, not forming a vascular plug, and a stomach which is simple and bulbous and leads directly into the pyloric duct.

Description

Tetrastemma vermiculus (Fig. 44H) is up to about 20 mm long and 0.6–0.8 mm wide. The head is typically oval in shape, flattened and wider than the adjacent body regions, with two pairs of cephalic furrows. The general body colour is dull whitish, salmon, pink, pale orange or, less frequently, a rich apricot yellow; the intestinal region occasionally appears pale green. On either side of the head a dark brown longitudinal pigment streak, characteristic for the species, extends between the anterior and posterior eyes. The streaks may be of uniform width or be broader near the front. A single mid-dorsal whitish stripe is sometimes present, commencing between the posterior pair of eyes; this stripe may be quite short or extend some distance down the body, although it is not always evident.

Often locally quite abundant, *Tetrastemma vermiculus* is a lively and restless species found on the lower shore and sublittorally to depths of about 40 m. It occurs under rocks and stones, on small algae or bryozoans, or in laminarian holdfasts. Knight-Jones & Nelson-Smith (1977) record the species as common in a *Halichondria* (sponge) assemblage where there are strong currents free from gravel. The species appears to have a prolonged reproductive period, as ripe individuals have been found from late autumn through to early summer.

The geographic range of *Tetrastemma vermiculus* extends from the British Isles (Fig. 47B) to Scandinavia, the Mediterranean, Madeira, and the eastern coast of North America from the Bay of Fundy southwards to the Gulf of Mexico.

Family DREPANOPHORIDAE

Genus *PUNNETTIA* Stiasny-Wijnhoff, 1926

Diagnosis

Benthic polystiliferous hoplonemerteans with a residual inner circular body wall muscle layer present in the cerebral and foregut regions; mouth and proboscis pore separate, mouth anterior to cerebral ganglia; rhynchocoel wall composed of a meshwork of interwoven longitudinal and circular muscle fibres; rhynchocoel with long unbranched lateral diverticula; head with large oblique cephalic furrows subdivided into secondary slits; nervous system with neither neurochords nor neurochord cells, without accessory lateral nerve; cerebral sensory organs large, alongside or behind cerebral ganglia, with two sensory canals and glandular appendage; eyes numerous, usually arranged in four longitudinal rows; blood vascular system comprising three longitudinal vessels with pseudometameric transverse connectives in intestinal region, mid-dorsal vessel entering rhynchocoel wall; excretory system in vicinity of cerebral ganglia and foregut, with ventral nephridiopores; intestinal caecum present; parenchymatous connective tissues well developed; sexes separate.

Punnettia splendida (Keferstein, 1862)
(Fig. 48)

Borlasia splendida Keferstein, 1862
Cerebratulus spectabilis McIntosh, 1869
Amphiporus spectabilis McIntosh, 1873–74, 1875a, b; Koehler, 1885; Tanner, 1908
Drepanophorus rubrostriatus McIntosh, 1875b; Riches, 1893; Sheldon, 1896; Allen, 1899
Drepanophorus spectabilis M.B.A., 1904, 1931, 1957; Wijnhoff, 1912; Southern, 1913; Campbell, 1976 (in part)
Drepanophoris rubrostriatus Tanner, 1908
Punnettia splendida Stiasny-Wijnhoff, 1934, 1936

Specific internal characters

Oesophagus long and with well developed longitudinal muscles; pyloric tube short; proboscis with 24–26 nerves.

Description

Up to 40–50 mm or more long and 4–5 mm wide, *Punnettia splendida* (Fig. 48A) typically has a flattened body with thin lateral margins, a rather broad and oar-like tail and a distinct spatulate head which is clearly delimited from the body by the cephalic furrows (Fig. 48B). These furrows are large and each contains seven or eight longitudinal secondary slits. There are about 70 eyes, arranged in four longitudinal rows on the dorsal side of the head.

Dorsally the colour is buff or reddish-brown marked with five longitudinal stripes of white, grey or pale pink; the lateral margins too are whitish. The brownish pigmentation between the margins and the outermost stripes does not extend on to the head, whereas the remaining dorsal pigmentation appears as four obvious dark cephalic lines. The tip and edges of the head are pale. Ventrally the body is an overall yellowish-pink or pinkish-brown without markings.

Punnettia splendida is mostly found sublittorally at depths down to 35–40 m amongst dredged algae, on coarse or fine gravelly sediments containing some mud or sand, or in honeycombed stones. It has also been found on oysters or in the tunicate *Ciona intestinalis* (Linnaeus). When irritated it will swim actively for a short while. Sexually mature females have been recovered in November.

In external appearance very similar to two Mediterranean drepanophorids, *Drepanophorus rubrostriatus* Hubrecht, 1875, and *Punnettia spectabilis* (Quatrefages, 1846), *Punnettia splendida* differs from both of these in its internal anatomy and, according to Stiasny-Wijnhoff (1934), is not found further south than the English Channel.

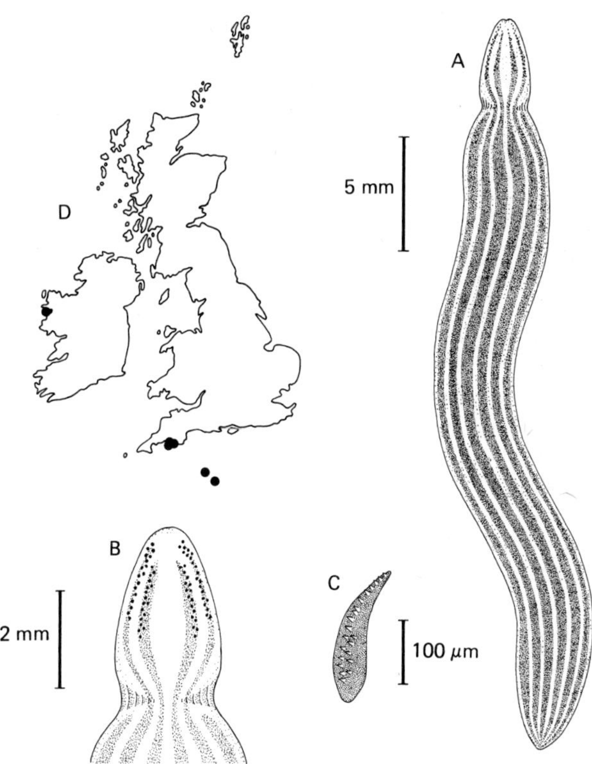

Fig. 48. *Punnettia splendida*; A, dorsal view of whole animal; B, enlarged view of head to show the distribution of the eyes; C, the stylet apparatus of the proboscis; D, recorded distribution from the British Isles. A, B redrawn from Bürger (1895) and Campbell (1976), C redrawn from Bürger (1895).

Family PARADREPANOPHORIDAE

Genus *PARADREPANOPHORUS* Stiasny-Wijnhoff, 1926

Diagnosis

Benthic polystiliferous hoplonemerteans with a strongly developed body wall inner circular muscle layer; mouth and proboscis pore close to each other but separate, sometimes opening via common atrial aperture anterior to cerebral ganglia; rhynchocoel wall composed of a meshwork of interwoven longitudinal and circular muscle fibres; rhynchocoel with long unbranched lateral diverticula; head with oblique cephalic furrows which may or may not be subdivided into secondary slits; nervous system with neither neurochords nor neurochord cells, without accessory lateral nerve; cerebral sensory organs large, alongside or behind cerebral ganglia, with one sensory canal and no glandular appendage; eyes numerous, usually arranged in four longitudinal rows; blood vascular system comprising three longitudinal vessels with pseudometameric transverse connectives in intestinal region, mid-dorsal vessel entering rhynchocoel wall; excretory system between cerebral organs and foregut, with ventral nephridiopores; intestinal caecum present; parenchymatous connective tissues well developed; sexes separate.

Paradrepanophorus crassus (Quatrefages, 1846)
(Fig. 49)

Cerebratulus crassus Quatrefages, 1846
Cerebratulus marginatus Renouf, 1931
Paradrepanophorus crassus Sheppard, 1935

Specific internal characters

Frontal organ absent; dorsoventral muscles abundant and strongly developed.

Description

A large and bulky species, *Paradrepanophorus crassus* (Fig. 49A) may be up to 16 cm long and 8–9 mm wide. The body is distinctly flattened and broadest in the mid-intestinal regions. Both the head and tail are bluntly pointed. The head is separated from the trunk by cephalic furrows which are characteristically white with brown secondary slits. Four dark longitudinal stripes on the dorsal cephalic surface mark the position of the eyes. The general colour is a light to dark brown dorsally, shading to dull orange near the pale lateral margins, light rose to pinkish-white on the ventral surface. Often the proboscis and intestinal diverticula are visible through the body wall.

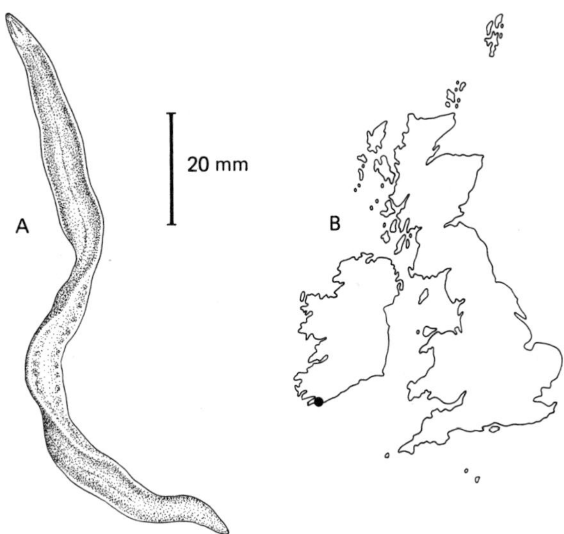

Fig. 49. *Paradrepanophorus crassus*; A, general view of whole animal; B, recorded distribution from the British Isles. A redrawn from Bürger (1895).

In British waters *Paradrepanophorus crassus* has been found under stones just below low water level. Sheppard (1935) notes that 'The mode of life of the worm is extremely interesting and very characteristic. It is always found lying in a delicate membranous, glistening, parchment-like tube, which is secreted by itself, fixed to the under side of large stones. It is almost invariably accompanied by a specimen of a small red Polychaete, *Staurocephalus rubrovittatus* Grube (now *Dorvillea rubrovittata*), which lives commensal with it.' On Mediterranean coasts it occurs at depths of 1–5 m in holes between worm-tubes and rocks.

Family MALACOBDELLIDAE

Genus *MALACOBDELLA* Blainville, 1827

Diagnosis

Nemerteans with broad, dorsoventrally compressed bodies bearing a single posterior ventral sucker; proboscis unarmed; rhynchocoel opening into barrel-shaped foregut which is lined by rows of motile, ciliated papillae; intestine sinuous, without lateral diverticula; eyes and cerebral sensory organs absent; sexes separate, gonads generally numerous, closely packed on either side of intestine; with entocommensal habits, typically living in the mantle chamber of bivalve molluscs, especially in species of the families Mactridae, Pholadidae and Veneridae.

Malacobdella grossa (Müller, 1776)
(Fig. 50)

Hirudo grossa Müller, 1776; Dalyell, 1853; Johnston, 1850
Phylline grossa Johnston, 1834
Hirudo anceps Dalyell, 1853
Malacobdella anceps Johnston, 1865
Malacobdella valenciennaei Johnston, 1865; Leslie & Herdman, 1881; Evans, 1909
Malacobdella grossa Montagu, 1808; Johnston, 1846, 1865; Parfitt, 1867; McIntosh, 1875c, 1927; Leslie & Herdman, 1881; Gibson, 1886; Riches, 1893; Sheldon, 1896; Gemmill, 1901; Maclaren, 1901; M.B.A., 1904, 1931, 1957; Evans, 1909; Jackson, 1935; Eales, 1952; Clark & Cowey, 1958; Bruce *et al.*, 1963; Gibson, 1967, 1968a, b; Gibson & Jennings, 1967, 1969; Jennings, 1968; Jennings & Gibson, 1968; Laverack & Blackler, 1974; Slinger & Gibson, 1974, 1975; Slinger, 1975; Jones *et al.*, 1979
Malacobdella sp. Dakin, 1909; Moore, 1937; Allen, 1969

Specific internal characters

Rhynchocoel full body length; nephridiopores of excretory system open ventrolaterally; lateral nerve cords unite above intestine just in front of anus; gonads numerous and closely packed.

Description

Malacobdella grossa (Fig. 50A) is a broad, flattened and leech-like species with a typically elongate oval shape. Sexually mature individuals reach lengths of 20–40 mm and are 8–15 mm wide. Immature worms are a translucent white or pale grey colour, with no evidence of gonads. In ripe males, however, the testes are a rich cream colour, often with a pink or rosy tint, and gravid females contain dark olive-green to deep yellowish-green ovaries. Occasional specimens may be dark brown, chocolate or almost black due to an infection with the sporozoan parasite *Haplosporidium malacobdellae* Jennings & Gibson, 1968. A heavy infection leads to parasitic castration of the nemerteans.

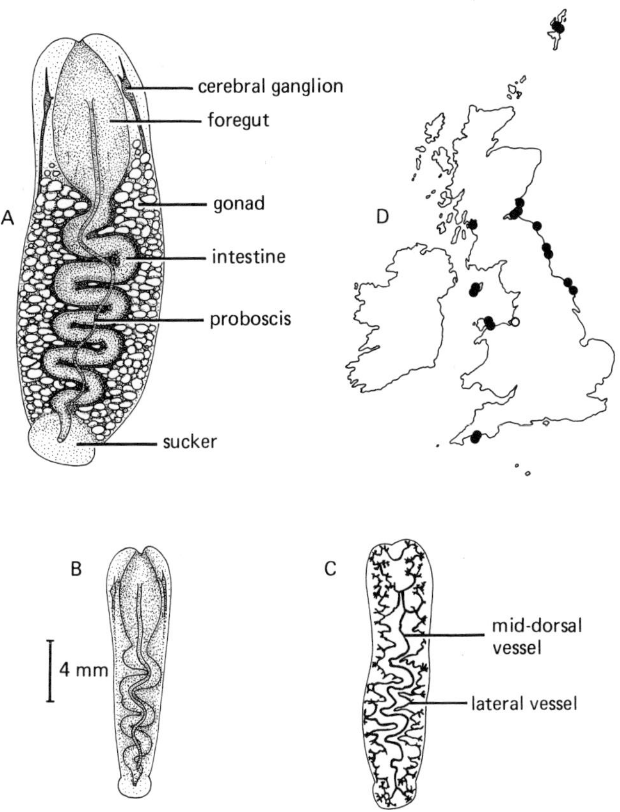

Fig. 50. *Malacobdella grossa*; A, drawing of mature adult stained to show the principal anatomical features; B, a juvenile individual; C, diagram of the blood vascular system; D, recorded distribution from the British Isles; ● = positive records, ○ = specific identity unknown. C redrawn from Gibson & Jennings (1967).

Malacobdella grossa is the only British representative of this unusual nemertean genus. It possesses a low degree of host specificity and has been recorded from 23 species of bivalve molluscs (Gibson, 1967; Jones *et al.*, 1979). British hosts reported are *Acanthocardia echinata* (Linnaeus), *Arctica islandica* (Linnaeus), *Cerastoderma edule* (Linnaeus), *Hiatella arctica* (Linnaeus), *Mya truncata* Linnaeus, *Pecten* sp., *Venus* (*Venus*) *casina* Linnaeus and *Zirfaea crispata* (Linnaeus). A solitary record exists of the nemertean being found in an unspecified fish (M.B.A., 1931). The maximum number of nemerteans obtained from a single host is five (Gibson, 1967) and such multiple infections are invariably composed of young individuals. Double infections are not uncommon, but more usually only a single worm occurs in any one host.

Malacobdella grossa is widely distributed on the coasts of Europe and on the Atlantic and Pacific coasts of North America.

Unidentified, undescribed and dubious species

Amongst the literature relating to British nemerteans four further categories of records can be recognised. These are references which merely report 'unidentified nemerteans', those which record unnamed species of the more common genera (e.g., *Amphiporus* sp., *Cerebratulus* sp., *Lineus* sp., *Tetrastemma* sp.), those referring to examples of valid genera for which there are no specifically named British representatives, and those including named species whose taxonomic status is particularly uncertain. It is with the last two categories that this section is concerned.

Unidentified species of valid genera

The three species belonging to this category are all monostiliferous hoplonemerteans.

Family OTOTYPHLONEMERTIDAE

Genus *OTOTYPHLONEMERTES* Diesing, 1863

Ototyphlonemertes sp.

Otyphlonemertes sp. Boaden, 1963

The genus *Ototyphlonemertes*, with a world-wide distribution, contains at least 24 species (Gibson, 1982) of which several occur in European waters. Members of this genus are all marine, interstitial forms inhabiting intertidal and shallow sublittoral sands and are typically small (3–50 mm long), extremely slender (less than 0.5 mm diameter), lack eyes and possess a single pair, rarely more, of statocysts housing statoliths in the cerebral ganglia (Kirsteuer, 1977). Boaden (1963) records an unidentified species as an occasional member of fairly fine shell gravel containing medium sand between low water neap and mid-tide levels at Traeth Bychan, Anglesey.

Family TETRASTEMMATIDAE

Genus *ARENONEMERTES* Friedrich, 1933

Arenonemertes sp.

Arenonemertes sp. Boaden, 1963

Boaden (1963), in a survey of the interstitial fauna of North Wales beaches, records as rare a species of *Arenonemertes* occurring in fine shell gravel with coarse sand between low water neap and mid-tide levels at Traeth Bychan, Anglesey. Included in the family Tetrastemmatidae by Gibson (1982), the genus comprises minute (1–3 mm long) mesopsammic sand-dwelling hoplonemerteans with four or fewer eyes, an intestine which lacks lateral diverticula but possesses three long anterior pouches, few gonads and large adhesive gland cells distributed over the entire body surface. The two described species, *Arenonemertes microps* Friedrich, 1933, and *Arenonemertes minutus* Friedrich, 1949, are both known so far only from the Baltic Sea.

Genus *PROSTOMATELLA* Friedrich, 1935

Prostomatella sp.

Prostomatella sp. Boyden & Little, 1973; Little & Boyden, 1976; Boyden *et al.*, 1977

This genus comprises species with small cephalic glands, cerebral sensory organs lying adjacent to the anterior margins of the cerebral ganglia and a short excretory system discharging to the exterior via several pairs of nephridiopores. There are two European species, *Prostomatella arenicola* Friedrich, 1935, an interstitial form dwelling in sand in the Baltic Sea, and *Prostomatella obscurum* (Schultze, 1851), widespread in Scandinavian waters and tolerant of reduced salinities. The British species occurs in clean sand at Minehead and Weston in the Severn estuary with densities up to $50/m^2$.

Dubious species

Ascaris flustrae Dalyell, 1853

Ascaris flustrae Dalyell, 1853
Cephalotrix? *flustrae* Johnston, 1865

A small, filiform worm found among decaying tissues of a polyzoan on the coast of Scotland, dark grey or brownish in colour with a darker anterior line and two distinct black eyes located on the clear tip of the head. Dalyell's original description and illustration are totally inadequate; whether this form belongs with the nemerteans, as suggested by Johnston (1865), or in some other phylum remains uncertain.

Gordius viridis Dalyell, 1853

Gordius viridis spinifer Dalyell, 1853
Stylus viridis Johnston, 1865

Recorded as a rare form on the Scottish coast, *Gordius viridis* is about 6–7 cm in overall length, including a slender and extremely retractile caudal cirrus which itself may be 1 cm or more long. The head bears distinct lateral cephalic furrows. In colour the body is a uniform 'mountain-green'. A long caudal cirrus and distinct horizontal cephalic furrows are characteristic features of several lineid heteronemertean genera but in the absence of additional anatomical data this nemertean cannot be further identified. As illustrated by Dalyell (1853) it superficially resembles a micrurid, but it cannot be related to any of the British *Micrura* species or those of other European waters.

Lineus linearis Montagu, 1808

Lineus linearis Montagu, 1808
Gordius linearis Montagu, 1808

In an unpublished notebook, Montagu briefly described long, slender, cream-coloured worms obtained from depths of 12–15 cm in sand near low tide level at Dawlish, Devonshire. He commented that although mostly 3–4 m in length, some examples were 5–6 m long. The worms produced copious amounts of mucus. This form appears neither to have been found subsequently nor more fully described elsewhere; although possibly a species of nemertean, its size and colour do not permit it to be even provisionally related to any of the existing British or European forms.

Meckelia asulcata McIntosh, 1873–74

Meckelia asulcata McIntosh, 1873–74; Riches, 1893; Sheldon, 1896

The generic name *Meckelia*, which is no longer valid, has been most commonly employed for species of the heteronemertean genera *Cerebratulus*, *Lineus* and *Micrura*; it has also been used for members of other groups, including *Amphiporus*, *Baseodiscus*, *Emplectonema* and *Tubulanus*. McIntosh's form is a heteronemertean which possesses neither cephalic furrows nor eyes, has a thick, rounded body with a *Lineus*-like body wall construction, and a proboscis containing two muscle layers. Up to 10–12 cm long, rose-pink in colour anteriorly but pale posteriorly, *Meckelia asulcata* has been found intertidally at and near St Magnus Bay, Shetland, and at Herm in the Channel Islands. Apart from the absence of lateral horizontal cephalic furrows, which exclude the form from any of the lineid heteronemertean genera, it is not possible to further identify McIntosh's species.

Micrura pardalis Haddon, 1886

Micrura pardalis Haddon, 1886b

Haddon (1886b) reported that 'The anterior portion of a Nemertean was dredged in Berehaven (south-western Ireland); it probably belongs to the genus *Micrura*. It was 35 mm long and 2 mm broad; flattened, head slightly swollen, gradually tapering anteriorly; trunk of body produced on the sides into a thin band, really forming a continuous lateral fin; no eyes were detected; a large lateral cephalic groove, distinctly red at the wide posterior end; colour dull pale, creamy orange, irregularly sprinkled with burnt sienna spots, which are more numerous at the anterior end; cephalic ganglia shine through with a marked rosy colour. If this species should prove to be new, the specific name of *pardalis* would be appropriate for it'.

The sharp lateral body margins and flattened shape described by Haddon are more typically associated with species of *Cerebratulus*, and the colour pattern somewhat resembles that of *Cerebratulus fuscus* (p. 76). No subsequent description was ever published, however, and the identity of this form must remain uncertain.

Nemertes assimilis Örsted, 1843

Nemertes assimilis Örsted, 1843; Evans, 1909

Originally found in Scandinavia, *Nemertes assimilis* is recorded by Evans (1909) from near Bass Rock in the North Sea. Information on this form is minimal; it is described as 35 mm long, 2 mm wide, with a cylindrical body bluntly rounded at either end, brownish-yellow in colour, being more brown anteriorly, and with 12 eyes. The generic name *Nemertes*, although mostly applied to *Emplectonema* species, has also been used for members of other genera (*Amphiporus*, *Lineus*, *Tetrastemma*). Which of these genera, if any, *Nemertes assimilis* belongs with is not known although the number of eyes excludes it from the tetrastemmid group.

Valencinia lineformis McIntosh, 1873–74

Valencinia lineformis McIntosh, 1873–74, 1875b; Riches, 1893; Sheldon, 1896

Valencinia lineiformis McIntosh, 1875a

McIntosh's (1873–74) description of *Valencinia lineformis* contains little information of either generic or specific value. 15–20 cm long, it resembles *Valencinia longirostris* (p. 110) in colour, in its general shape and in lacking cephalic furrows. It differs significantly, however, in possessing a row of small eyes on either side of the head and in having the lateral nerve cords running in the body wall outer longitudinal muscle layer. The presence of eyes prevents this form from being regarded as synonymous with *Valencinia longirostris* (as Bürger, 1895, 1904, did) and because of the inadequate description the taxonomic status of McIntosh's species remains uncertain.

Valencinia lineformis was obtained from amongst shelly gravel and on the red alga *Corallina officinalis* Linnaeus from about 10 m depth at Bressay Sound, Shetland Islands.

Vermiculus lineatus Dalyell, 1853

Vermiculus lineatus Dalyell, 1853

Cephalotrix lineatus Johnston, 1865

Dalyell (1853) described *Vermiculus lineatus* as '. . . very slender. Anterior extremity obtuse, with two black eyes on the surface, near the front. Posterior extremity tapering. Colour universally dark grey, with a white line down the back; anterior extremity, wherein the eyes are seated, white. Motion smooth and gliding.'

'A smaller specimen, with similar eyes, but the anterior portion ruddy, I conjectured might be a young animal of the same species'.

Mainly used for species of the hoplonemertean genera *Amphiporus*, *Oerstedia* and *Tetrastemma*, the name *Vermiculus* has also been employed for members of the heteronemertean genus *Lineus* and, indeed, McIntosh (1873–74) lists *Vermiculus lineatus* as a synonym of *Lineus gesserensis* (see p. 90), commenting that Dalyell's figure depicts a young example with only two eyes. Dalyell's description of the form, however, leaves McIntosh's conclusion open to doubt. Whilst neither the generic nor specific identity of *Vermiculus lineatus* can be determined, its colour resembles varieties described for both *Amphiporus lactifloreus* (p. 120) and *Oerstedia dorsalis* (p. 144). The number of eyes, conversely, is at variance with both of these forms.

Glossary

accessory lateral nerve A longitudinal nerve, whose origin lies in the dorsal cerebral lobes, closely associated with and running dorsal to the main lateral nerve cords; found only in hoplonemerteans, typically in members of the family Prosorhochmidae.

accessory pouch A chamber in the proboscis of hoplonemerteans, located just in front of the central stylet armature, which houses replacement or accessory stylets; there are typically two such pouches in a proboscis, although more are found in a few species.

accessory stylet A replacement stylet, grown and housed in the accessory pouches of the proboscis in hoplonemerteans.

atrial chamber A subterminal anterior chamber occurring in some hoplonemerteans, particularly in the suborder Polystilifera, into which the alimentary canal and proboscis open separately; the aperture of the atrium is thus common to both the gut and proboscis.

basement membrane A slender connective tissue lining below the epidermis, occasionally quite well developed.

caecum Blind-ending anterior ventral extension of the intestine, typically found in hoplonemerteans, often with anterior and/or lateral diverticula.

caudal cirrus A small, tail-like extension at the posterior end of the body, typically found in certain heteronemertean genera.

central stylet A needle-like structure forming the proboscis armature of hoplonemerteans, mounted on a stylet basis; in the suborder Monostilifera there is only a single central stylet, but in the Polystilifera the armature consists of several minute stylets attached to a shield- or pad-like basis.

cephalic Appertaining to the anterior (head) part of the body.

cephalic furrow, groove or slit A longitudinal, oblique or transverse depression in the cephalic epidermis, sometimes very deep, lined by a ciliated epithelium which normally lacks gland cells; the furrows are paired and their distribution on the head may be of taxonomic importance; typically, although not always, the cerebral sensory organs open into the furrows.

cephalic glands Mucus-producing glands located in the anterior region of the head, in some species enormously developed and posteriorly extending beyond the cerebral ganglia; generally the glands either discharge via the frontal organ, when one is present, or through numerous improvised ducts which penetrate the cephalic epidermis.

cephalic lobe The rounded, semi-circular, heart-shaped or flattened anterior tip of the body, clearly distinguishable from the trunk, which occurs in many species of nemerteans; although it quite often bears eyes and cephalic furrows, the cephalic lobe is not strictly equivalent to a head as it usually does not contain the cerebral ganglia.

cerebral canal A ciliated tubular duct connecting the cerebral sensory organs with the body surface; often opening into the cephalic furrows.

cerebral ganglia The brain lobes, characteristically consisting of dorsal and ventral pairs of cerebral lobes, in close connection with each other, linked by transverse commissures; the rhynchocoel passes between the cerebral ganglia.

cerebral lobe Part of the cerebral ganglia, comprising a central fibrous region surrounded by a ganglionic cell layer; dorsal and ventral lobes on each side of the body are closely joined together and may be virtually indistinguishable.

cerebral sensory organ A sensory organ, whose functions are not fully understood, consisting of a complex of nervous and glandular tissues surrounding a ciliated canal; in heteronemerteans the organs are closely attached to the posterior margins of the dorsal cerebral lobes, in other groups they form discrete structures, sometimes positioned some distance in front of the cerebral ganglia but linked to them by nerves; cerebral sensory organs are missing from some palaeonemerteans, many polystiliferous hoplonemerteans and the genera *Carcinonemertes* and *Malacobdella*.

connective tissue A tissue consisting of a mixture of mucopolysaccharide ground substance, proteinaceous filaments and free cells of various types, typically forming the general membranes of the body which surround blood vessels, nerves, excretory tubules and other structures, or serving as the packing tissue filling the space between the body wall and the various organ systems.

dermis The tissue layer between the epidermis and the body wall musculature; usually comprising a mixture of connective tissues and gland cells, best developed in the heteronemerteans; the organisation of the dermis is of taxonomic significance.

diverticulum A lateral or anterior pouch projecting from the alimentary canal or rhynchocoel; lateral diverticula usually occur in pairs and those of the intestine typically exhibit pseudometamerism.

dorsal commissure The transverse nerve connective joining the dorsal cerebral lobes; the rhynchocoel passes below the dorsal commissure.

efferent tubules Those parts of the excretory system which lead from the general collecting ducts to the nephridiopores.

entocommensal A commensal organism living in a semi-enclosed chamber or cavity of a host species which is in open communication with the external environment but which affords a sheltered habitat.

epithelial barbs Minute needle-like structures contained in the proboscis epithelium of several palaeo- and heteronemertean species, usually arranged in tightly packed groups; the function of the barbs appears to be to assist the proboscis to grip prey organisms.

flame cell Proximal terminal portion of the excretory system, so called because the beating of its long cilia creates an illusion of a flickering flame.

frontal organ A flask-shaped epidermal pit situated on the tip of the head, lined by a ciliated but non-glandular epithelium, best developed in hoplonemerteans; in several heteronemertean species the single frontal organ is replaced by three histologically similar structures which may open independently or through a common pore.

ganglionic cells Nerve cells of various types which typically form an ensheathing layer around the cerebral ganglia and lateral nerve cords.

hermaphroditism The condition whereby both male and female gametes are produced by the same individual.

inner neurilemma A connective tissue layer separating the fibrous tissues and ganglionic cells of the cerebral lobes.

lacunae Spacious blood channels which lack muscle fibres in their walls, often subdivided by connective tissue strands.

lateral blood vessels The main pair of longitudinal blood vessels extending the full length of the body in a lateral or ventrolateral position; invariably forming distinct thick-walled vessels in the intestinal regions, they are sometimes developed into lacunae in the vicinity of the foregut.

lateral nerve cords The main pair of longitudinal nerves leading posteriorly from the rear of the ventral ganglionic lobes, running laterally or ventrolaterally; the position of the lateral nerve cords relative to the various body wall layers is of major taxonomic importance.

mid-dorsal blood vessel A blood vessel, invariably thick-walled, extending the full body length behind the cerebral ganglia, running just above the alimentary canal; typically the vessel enters the ventral rhynchocoel wall near the cerebral ganglia to form either a rhynchocoelic villus or a vascular plug; in some hoplonemerteans the vessel branches for a short distance and two vascular plugs are developed; in several palaeonemertean and polystiliferous hoplonemertean genera the mid-dorsal vessel is either reduced or missing.

muscle crosses Muscle fibres traversing muscle layers in the body wall (palaeonemerteans) or proboscis (heteronemerteans) to form a usually distinct cross-over distinguishable in transverse sections; the number of proboscis muscle crosses is of taxonomic significance.

nephridiopore The external opening of the excretory system; typically there are one to a few pairs of nephridiopores, but increased development of the excretory system in terrestrial and freshwater nemerteans has resulted in these forms possessing very large numbers of nephridiopores; unusually, but a characteristic feature of several baseodiscid species, internal nephridiopores are developed which discharge into the foregut lumen.

neurochord cells Very large and obvious nerve cells situated amongst the ganglionic cells of the cerebral ganglia; the presence or absence of neurochord cells is of taxonomic significance.

neurochords Longitudinal extensions or processes leading into the lateral nerve cords from neurochord cells.

neuroglandular An association of nervous and gland cells in a specific organ.

outer neurilemma A connective tissue layer investing the outer surface of the cerebral ganglia and lateral nerve cords.

oviparous Laying eggs; fertilisation external, although accidental internal fertilisation may occur irregularly just before the eggs are laid; by far the commonest form of sexual reproduction by nemerteans.

ovoviviparous Egg development after internal fertilisation leading to the formation of young worms taking place entirely within the ovaries, but the developing young nourished by the egg yolk stores only and not through tissue involvement of the maternal organs; very uncommon in nemerteans.

parenchyma The body packing tissue, filling the space between the body wall layers and the internal organs, composed of connective tissues.

proboscis An eversible, muscular, tubular organ housed, when retracted, in the rhynchocoel; the organisation of the proboscis is of major taxonomic importance.

proboscis pore The aperture through which the proboscis is everted; situated terminally or subterminally on the head, the proboscis pore is separate in anoplan nemerteans and in some hoplonemerteans but is generally shared by the gut in enoplan forms.

protandrous The condition whereby a hermaphroditic organism is functionally a male first, a female later.

pseudometamerism The serial replication of body structures, in nemerteans typically the intestinal diverticula, gonads and transverse blood vessels joining the lateral and mid-dorsal vessels, without the formation of segments; confined in nemerteans to the intestinal regions of the body.

pylorus A slender and often elongate posterior extension of the foregut, joining the stomach with the intestine and usually opening into the dorsal intestinal wall.

rhynchocoel The dorsal fluid-filled chamber housing the proboscis, situated above the alimentary tract.

rhynchocoelic villus Typically found in heteronemerteans, a thin-walled elongate protrusion of the mid-dorsal blood vessel into the rhynchocoel lumen.

rhynchodaeum A short muscular chamber located in the head and opening at the proboscis pore; the junction between rhynchodaeum and rhynchocoel marks the insertion of the proboscis.

sexual dimorphism The condition when males and females of the same species exhibit external differences which, in British nemerteans, are restricted to colour variations.

splanchnic muscles Longitudinal, oblique or circular muscle fibres specifically associated with the foregut, most commonly found in the heteronemerteans.

stylet basis A cylindrical (Monostilifera) or pad- to shield-shaped (Polystilifera) structure on which the stylet armature of the hoplonemertean proboscis is mounted.

stylet bulb region The middle portion of the hoplonemertean proboscis, housing the stylet armature.

submuscular capillary network A profusely branching system of small thin-walled blood vessels located just internal to the body wall musculature; typically associated with certain species of terrestrial nemerteans.

transverse connectives Pseudometamerically arranged transverse blood vessels which link the mid-dorsal and lateral vessels in the intestinal regions of the body; transverse connectives are missing from palaeonemerteans and several enoplan nemerteans.

vascular plug Typically associated with hoplonemerteans, a bulbous protrusion of the mid-dorsal blood vessel through the ventral rhynchocoel wall, usually located close to the cerebral ganglia.

vas deferens A longitudinal duct, discharging into the rear of the intestine in male *Carcinonemertes*, into which the vasa efferentia open.

vas efferens A short tubular duct leading from a testis to the vas deferens in male *Carcinonemertes.*

ventral commissure The transverse nerve connective joining the ventral cerebral lobes; the rhynchocoel passes above the ventral commissure.

Acknowledgements

I am most grateful to Dr J. Moore, Department of Zoology, University of Cambridge, Professor M. S. Laverack, Gatty Marine Laboratory, University of St Andrews, Drs G. Berg and P. Sundberg, Department of Zoology, University of Göteborg, Dr R. Anderson, Department of Agricultural and Food Chemistry, Queen's University, Belfast, Mr T. Venn, Millport Marine Biological Station, and Dr I. M. Varndell, Department of Histochemistry, Hammersmith Hospital, London, for their various comments, criticisms, suggestions and unpublished records. In particular their field testing of the key has been most helpful and in this area, and for additional data, I also wish to thank Dr R. S. K. Barnes, Department of Zoology, University of Cambridge. Amongst my own students who laboured to identify material through earlier copies of the key I should like especially to thank Eric McEvoy, Clare McNicholas and Gillian Wright.

The conscientious efforts of Jacqueline Harrison and Jane Glover in helping to trace much of the literature are greatly appreciated. Parts of the text are based upon practical studies supported by Science Research Council grant numbers B/RG–87945 and GR/A–86350, whose assistance is gratefully acknowledged.

References

References marked * deal with the broader aspects of nemertean biology and are of general interest, that denoted with † has not been seen in the original.

† Abildgaard, P. C. 1806. Article on *Planaria dorsalis*. In *Zoologica Danica*, vol. 4, ed. O. F. Müller.

Allen, E. J. 1899. On the fauna and bottom-deposits near the thirty-fathom line from the Eddystone Grounds to Start Point. *J. mar. biol. Ass. U.K.* **5,** 365–542.

Allen, E. J. & Todd, R. A. 1900. The fauna of the Salcombe Estuary. *J. mar. biol. Ass. U.K.* **6,** 151–217.

Allen, J. A. 1969. Observations on size composition and breeding of Northumberland populations of *Zirphaea crispata* (Pholadidae: Bivalvia). *Mar. Biol.* **3,** 269–75.

Anderson, R. 1980. The status of the land nemertine *Argonemertes dendyi* (Dakin) in Ireland. *Ir. Nat. J.* **20,** 153–7.

Anonymous, 1854. *The North British Review*, **22,** 1–56.

Barrett, J. H. & Yonge, C. M. 1958. *Collins Pocket Guide to the Sea Shore.* London, Collins, 272 pp.

Bartsch, I. 1973. Zur Nahrungsaufnahme von *Tetrastemma melanocephalum* (Nemertini). *Helgoländer wiss. Meeresunters.* **25,** 326–31.

Bassindale, R. 1941. Studies on the biology of the Bristol Channel. IV. The invertebrate fauna of the southern shores of the Bristol Channel and Severn Estuary. *Proc. Bristol Nat. Soc.* **9,** 143–201.

Bassindale, R. 1943. Studies on the biology of the Bristol Channel. XI. The physical environment and intertidal fauna of the southern shores of the Bristol Channel and Severn Estuary. *J. Ecol.* **31,** 1–29.

Bassindale, R. & Barrett, J. H. 1957. The Dale Fort marine fauna. *Proc. Bristol Nat. Soc.* **29,** 227–328.

Beattie, W. 1858. On the reproduction of *Nemertes borlassii. Proc. zool. Soc. Lond.* **26,** 307.

Beaumont, W. I. 1895a. Report on nemertines observed at Port Erin in 1894 and 1895. *Proc. Trans. Lpool biol. Soc.* **9,** 354–73.

Beaumont, W. I. 1895b. Report on nemertines observed at Port Erin in 1894 and 1895. *Lpool mar. biol. Comm. Rep.* **4,** 449–68.

Beaumont, W. I. 1900a. The fauna and flora of Valencia Harbour on the west coast of Ireland. Part II – The benthos (dredging and shore-collecting). VII. Report on the results of dredging and shore-collecting. *Proc. R. Ir. Acad.* **5** (3), 754–98.

Beaumont, W. I. 1900b. The fauna and flora of Valencia Harbour on the west coast of Ireland. XI. Report on the Nemertea. *Proc. R. Ir. Acad.* **5** (3), 815–31.

Benham, W. B. 1892. Note on the occurrence of a freshwater nemertine in England. *Nature, Lond.* **46,** 611–12.

Benham, W. B. 1897. Fission in nemertines. *Q. Jl microsc. Sci.* **39,** 19–31.

Berg, G. 1972a. Studies on *Nipponnemertes* Friedrich, 1968 (Nemertini, Hoplonemertini). *Zool. Scr.* **1,** 211–25.

Berg, G. 1972b. Taxonomy of *Amphiporus lactifloreus* (Johnston, 1828) and *Amphiporus dissimulans* Riches, 1893 (Nemertini, Hoplonemertini). *Astarte*, **5,** 19–26.
Berg, G. 1973. On morphology and distribution of some hoplonemertean species along Scandinavian coasts (Nemertini). *Zool. Scr.* **2,** 63–70.
Bergendal, D. 1902. Einige Bemerkungen über *Carinoma armandi* Oudemans (sp. McInt.). *Öfv. K. Vet.-Akad. Förh., Stockholm*, **1,** 13–18.
Bergendal, D. 1903. Über 'Sinnesgrübchen' im Epithel des Vorderkopfes bei *Carinoma armandi* sp. McInt. (Oudemans) nebst einigen systematischen Bemerkungen über die Arten dieser Gattung. *Zool Anz.* **26,** 608–19.
Blainville, H. M. De, 1827. *Dictionnaire des Sciences Naturelles*, vol. 47, ed. F. G. Levrault, Paris, pp. 270–71.
Boaden, P. J. S. 1963. The interstitial fauna of some North Wales beaches. *J. mar. biol. Ass. U.K.* **43,** 79–96.
Böhmig, L. 1892. Report in Bericht der II Section für Zoologie. *Mitt. naturw. Ver. Steierm.* lxxxii–lxxxv.
Borlase, W. 1758. *The Natural History of Cornwall.* Jackson, Oxford, 326 pp.
Boyden, C. R., Crothers, J. H., Little, C. & Mettam, C. 1977. The intertidal invertebrate fauna of the Severn Estuary. *Fld Stud.* **4,** 477–554.
Boyden, C. R. & Little, C. 1973. Faunal distributions in soft sediments of the Severn Estuary. *Estuar. Coastal mar. Sci.* **1,** 203–23.
Braithwaite, R. R. & Clayton, E. B. 1945. A British freshwater nemertine. *Nature, Lond.* **156,** 237.
Bruce, J. R. 1948. Additions to faunal records, 1941–46. *Rep. mar. biol. Stn. Port Erin*, 39–58.
Bruce, J. R., Colman, J. S. & Jones, N. S. 1963. *Marine Fauna of the Isle of Man.* University Press, Liverpool, 307 pp.
Brunberg, L. 1964. On the nemertean fauna of Danish waters. *Ophelia*, **1,** 77–111.
Bürger, O. 1892. Zur Systematik der Nemertinenfauna des Golfs von Neapel. *Nachr. König. Ges. wiss. Georg-Augusts Univ. Göttingen*, **5,** 137–78.
*Bürger, O. 1895. Die Nemertinen des Golfes von Neapel und der Angrenzenden Meeres-Abschnitte. *Fauna Flora Golf. Neapel*, **22,** 1–743.
Bürger, O. 1904. Nemertini. *Das Tierreich*, **20,** 1–151.
Byerley, I. 1854a. The fauna of Liverpool. *Proc lit. phil. Soc. Lpool*, **8,** Appendix, 1–125.
Byerley, I. 1854b. *The Fauna of Liverpool.* Deighton & Laughton, Liverpool, 125 pp.
Campbell, A. C. 1976. *The Hamlyn Guide to the Seashore and Shallow Seas of Britain and Europe.* Hamlyn, London, 320 pp.
Cantell, C.-E. 1966. The devouring of the larval tissues during the metamorphosis of pilidium larvae (Nemertini). *Ark. Zool.* **18** (2), 489–92.
Cantell, C.-E. 1975. Anatomy, taxonomy, and biology of some Scandinavian heteronemertines of the genera *Lineus*, *Micrura*, and *Cerebratulus*. *Sarsia*, **58,** 89–122.
Cantell, C.-E. 1976. Complementary description of the morphology of *Lineus longissimus* (Gunnerus, 1770) with some remarks on the cutis layer in heteronemertines. *Zool. Scr.* **5,** 117–20.
Chumley, J. 1918. *The Fauna of the Clyde Sea Area.* University Press, Glasgow, 200 pp.
Claparède, E. 1862. Études anatomiques sur les annélides, turbellariés, opalines et grégarines observés dans les Hébrides. *Mém. Soc. Phys. Hist. nat. Genève*, **16,** 71–164.
Clark, R. B. & Cowey, J. B. 1958. Factors controlling the change of shape of certain nemertean and turbellarian worms. *J. exp. Biol.* **35,** 731–48.

Cloudsley-Thompson, J. L. & Sankey, J. 1961. *Land Invertebrates. A Guide to British Worms, Molluscs and Arthropods (excluding Insects).* Methuen, London, 156 pp.

Coe, W. R. 1902. The nemertean parasites of crabs. *Am. Nat.* **36,** 431–50.

*Coe, W. R. 1905. Nemerteans of the west and northwest coasts of America. *Bull. Mus. comp. Zool. Harv.* **47,** 1–318

Coe, W. R. 1930. Asexual reproduction in nemerteans. *Physiol. Zoöl.* **3,** 297–308.

Coe, W. R. 1931. A new species of nemertean (*Lineus vegetus*) with asexual reproduction. *Zool. Anz.* **94,** 54–60.

Coe, W. R. 1932. Regeneration in nemerteans. III. Regeneration in *Lineus pictifrons. J. exp. Zool.* **61,** 29–43.

Coe, W. R. 1940. Revision of the nemertean fauna of the Pacific coasts of North, Central, and northern South America. *Allan Hancock Pacif. Exped.* **2,** 247–323.

*Coe, W. R. 1943. Biology of the nemerteans of the Atlantic coast of North America. *Trans. Conn. Acad. Arts Sci.* **35,** 129–328.

Coe, W. R. 1944. Geographical distribution of the nemerteans of the Pacific coast of North America, with descriptions of two new species. *J. Wash. Acad. Sci.* **34,** 27–32.

Colgan, N. 1916. Observations on phototropism and the development of eye-spots in the marine nemertine, *Lineus gesserensis. Ir. Nat.* **25,** 7–12.

Corlett, J. 1947. Studies on the sedentary marine fauna of the Mersey Estuary. Master of Science Thesis, University of Liverpool, 79 pp.

Corrêa, D. D. 1958. Nemertinos do litoral Brasileiro (VII). *Anais Acad. bras. Cienc.* **29,** 441–55.

Corrêa, D. D. 1961. Nemerteans from Florida and Virgin Islands. *Bull. mar. Sci. Gulf Caribb.* **11,** 1–44.

Corrêa, D. D. 1963. Nemerteans from Curaçao. *Stud. Fauna Curaçao*, **17,** 41–56.

Cowey, J. B. 1952. The structure and function of the basement membrane muscle system in *Amphiporus lactifloreus* (Nemertea). *Q. Jl microsc. Sci.* **93,** 1–15.

Crothers, J. H. 1966. Dale Fort marine fauna. *Fld Stud.* **2,** Supplement, 1–169.

Dakin, W. J. 1909. The Marine Biological Station at Port Erin, being the twenty-third annual report of the Liverpool Marine Biology Committee. *Trans. biol. Soc. Lpool*, **23,** 1–62.

Dakin, W. J. 1915. Fauna of West Australia – III. A new nemertean, *Geonemertes dendyi*, sp. n., being the first recorded land nemertean from Western Australia. *Proc. zool. Soc. Lond.* 567–70.

Dalyell, J. G. 1853. *The Powers of the Creator Displayed in the Creation*, vol. 2. Van Voorst, London, 327 pp.

Danielli, J. F. & Pantin, C. F. A. 1950. Alkaline phosphatase in protonephridia of terrestrial nemertines and planarians. *Q. Jl microsc. Sci.* **91,** 209–13.

Davies, H. 1815. Some observations on the Sea Long-worm of Borlase, *Gordius marinus* of Montagu. *Trans. Linn. Soc. Lond.* **11,** 292–95.

Davies, H. 1816. Some observations on the Sea Long-worm of Borlase, *Gordius marinus* of Montagu. *Lond. Med. Phys. J.* **36,** 207–9.

Davies, H. 1817. Einige Bemerkungen über den See-Langwurm von Borlase, *Gordius marinus* von Montagu. *Isis*, 1054–56.

Delle Chiaje, S. 1825. *Memorie Sulla Storia e Notomia Degli Animali Senza Vertebre del Regno di Napoli*, vol. 2, 185–444. Societa' Tipografica, Napoli.

Delle Chiaje, S. 1841. *Descrizione et Notomia Degli Animali Invertebrati Della Sicilia Citeriore*, **3,** 125–30, Naples.

Diesing, K. M. 1850. *Systema Helminthum*, vol. 1, 238–77, Vindobonae.

Diesing, K. M. 1863. Nachträge zur Revision der Turbellarien. *Sitz. ber. Math. Nat. Kl. Akad. Wiss. Wien*, **46,** 173–88.

Dollfus, R. P. 1924. Contribution à la faune des invertébrés du banc de Rockall. *Bull. Inst. océanogr. Monaco*, No. 438, 1–28.
Duerden, J. E. 1894. Notes on the marine invertebrates of Rush, County Dublin. *Ir. Nat.* **3,** 230–3.
Dugès, A. 1828. Recherches sur l'organisation et les moeurs des Planariées. *Annls Sci. nat.* **15** (1), 139–83.
Durham, M. J. 1948. *Geonemertes dendyi* Dakin. *Rep. Lancs. Chesh. Fauna Comm.* **27,** 35.
Durham, M. J. 1949. Occurrence of the land nemertine, *Geonemertes dendyi* Dakin, in Wirral, Cheshire. *NWest. Nat.* **22,** 182–3.
Eagle, R. A. 1973. Benthic studies in the south east of Liverpool Bay. *Estuar. Coastal mar. Sci.* **1,** 285–99.
Eales, N. B. 1952. *The Littoral Fauna of Great Britain.* Cambridge University Press, 2nd edit. 305 pp.
Eason, D. G. 1973. Investigations on the ecology of the common British heteronemertean *Lineus ruber* (O. F. Müller). Master of Philosophy Thesis, Liverpool Polytechnic, 141 pp.
Eggers, F. 1935. Zur Bewegungsphysiologie von *Malacobdella grossa* Müll. *Z. wiss. Zool.* **147,** 101–31.
Ehrenberg, C. G. 1831. *Phytozoa turbellaria.* In P. C. Hemprich & C. G. Ehrenberg, *Symbolae Physicae, seu Icones et Descriptiones Corporum Naturalium Novorum aut Minus Cognitorum*, Berolini.
Elmhirst, R. 1922. Notes on the breeding and growth of marine animals in the Clyde sea area. *Rep. Scott. mar. biol. Ass.* 19–43.
Evans, R. G. 1949. The intertidal ecology of rocky shores in south Pembrokeshire. *J. Ecol.* **37,** 120–39.
Evans, W. 1909. Our present knowledge of the fauna of the Forth area. *Proc. R. phys. Soc.* **17,** 1–64d.
Evans, W. 1915. Two additions to the list of 'Forth' Nemertinea. *Scott. Nat.* 336.
Fabricius, O. 1780. *Fauna Groenlandica.* Hafniae & Lipsiae, 452 pp.
Farran, G. P. 1915. Results of a biological survey of Blacksod Bay, Co. Mayo. *Fish. Res., Dep. Agric. Ireland*, No. 3, 1–72.
Friedrich, H. 1933. Morphologische Studien an Nemertinen der Kieler Bucht, I und II. *Z. wiss. Zool.* **144,** 496–509.
Friedrich, H. 1935. Studien zur Morphologie, Systematik und Ökologie der Nemertinen der Kieler Bucht. *Arch. Naturgesch.* **4,** 293–375.
Friedrich, H. 1949. Über zwei bemerkenswerte neue Nemertinen der Sandfauna. *Kieler Meeresforsch.* **6,** 68–72.
Friedrich, H. 1955. Beiträge zu einer Synopsis der Gattungen der Nemertini monostilifera nebst Bestimmungsschlüssel. *Z. wiss. Zool.* **158,** 133–92.
Friedrich, H. 1968. *Sagaminemertes*, eine bemerkenswerte neue Gattung der Hoplonemertinen und ihre systematische Stellung. *Zool. Anz.* **180,** 33–6.
Gamble, F. W. 1896. Notes on a zoological expedition to Valencia Island, Co. Kerry. Shore-collecting and dredging. *Ir. Nat.* **5,** 129–36.
Gemmill, J. F. 1901. Marine worms. In *Fauna, Flora and Geology of the Clyde Area*, ed. G. F. S. Elliot, M. Laurie & J. B. Murdoch, Glasgow, pp 359–63.
Gibson, R. 1967. Occurrence of the entocommensal rhynchocoelan, *Malacobdella grossa*, in the Oval Piddock, *Zirfaea crispata*, on the Yorkshire coast. *J. mar. biol. Ass. U.K.* **47,** 301–17.
Gibson, R. 1968a. Studies on nutrition in the phylum Rhynchocoela, with observations on the ecology of one entocommensal species. Doctor of Philosophy Thesis, University of Leeds, 310 pp.
Gibson, R. 1968b. Studies on the biology of the entocommensal rhynchocoelan *Malacobdella grossa*. *J. mar. biol. Ass. U.K.* **48,** 637–56.

*Gibson, R. 1972. *Nemerteans*. Hutchinson, London, 224 pp.

Gibson, R. 1979. Nemerteans of the Great Barrier Reef. 2. Anopla Heteronemertea (Baseodiscidae). *Zool. J. Linn. Soc.* **66,** 137–60.

Gibson, R. 1982. Nemertea. In *Synopsis and Classification of Living Organisms*, ed. S. P. Parker, McGraw-Hill, New York **1,** 823–46.

Gibson, R. & Jennings, J. B. 1967. 'Leucine aminopeptidase' activity in the blood system of rhynchocoelan worms. *Comp. Biochem. Physiol.* **23,** 645–51.

Gibson, R. & Jennings, J. B. 1969. Observations on the diet, feeding mechanisms, digestion and food reserves of the entocommensal rhynchocoelan *Malacobdella grossa*. *J. mar. biol. Ass. U.K.* **49,** 17–32.

Gibson, R. & Moore, J. 1976. Freshwater nemerteans. *Zool. J. Linn. Soc.* **58,** 177–218.

Gibson, R. & Young, J. O. 1971. *Prostoma jenningsi* sp. nov. a new British freshwater hoplonemertean. *Freshwat. Biol.* **1,** 121–7.

Gibson, R. & Young, J. O. 1976. Ecological observations on a population of the freshwater hoplonemertean *Prostoma jenningsi* Gibson & Young 1971. *Arch. Hydrobiol.* **78,** 42–50.

Gibson, R. J. H. 1886. Report on the Vermes of the Liverpool Marine Biology Committee district. *The First Report upon the Fauna of Liverpool Bay and the Neighbouring Seas*, ed. W. A. Herdman, pp. 144–60.

Gontcharoff, M. 1951. Biologie de la régénération et de la reproduction chez quelques Lineidae de France. *Annls Sci. nat., Zool.* **13** (11), 149–235.

Gontcharoff, M. 1959. Rearing of certain nemerteans (genus *Lineus*). *Ann. N.Y. Acad. Sci.* **77,** 93–5.

Gontcharoff, M. 1960. Le développement post-embryonnaire et la croissance chez *Lineus ruber* et *Lineus viridis* (Némertes Lineidae). *Annls Sci. nat., Zool.* **2** (12), 225–79.

Goodsir, H. D. S. 1845. Descriptions of some gigantic forms of invertebrate animals from the coast of Scotland. *Ann. Mag. nat. Hist.* **15** (1), 337–83.

Graff, L. Von, 1879. *Geonemertes chalicophora*, eine neue Landnemertine. *Morph. Jb.* **5,** 430–49.

Gray, J. E. 1857. Notice of a large species of *Lineus*?, taken on the coast near Montrose. *Proc. zool. Soc. Lond.* **25,** 210.

Gray, J. E. 1858. Notice of a large species of *Lineus*? taken on the coast near Montrose. *Ann. Mag. nat. Hist.* **1** (3), 160.

Green, J. 1968. *The Biology of Estuarine Animals*. Sidgwick & Jackson, London, 401 pp.

Griffith, E. 1834. *The Animal Kingdom Arranged in Conformity with its Organization, by the Baron Cuvier with Supplementary Additions to each Order*, vol. 12, Whittaker, London, 601 pp.

Grube, E. 1855. Bemerkungen über einige Helminthen und Meerwürmer. *Arch. Naturgesch.* 137–58.

Gunnerus, J. E. 1770. Nogle smaa rare og meestendeelen nye norske søedyr. *Skr. Kbh. Selsk.* **10,** 166–76.

Haddon, A. C. 1886a. Preliminary report on the fauna of Dublin Bay. *Proc. R. Ir. Acad.* **4** (2), 523–31.

Haddon, A. C. 1886b. First report on the marine fauna of the south-west of Ireland. *Proc. R. Ir. Acad.* **4** (2), 599–638.

Herdman, W. A. 1894. Seventh annual report of the Liverpool Marine Biology Committee and their Biological Station at Port Erin. *Trans. biol. Soc. Lpool*, **8,** 1–55.

Herdman, W. A. 1896. Ninth annual report of the Liverpool Marine Biology Committee and their Biological Station at Port Erin. *Trans. biol. Soc. Lpool*, **10,** 1–59.

Herdman, W. A. 1900. The fourteenth annual report of the Liverpool Marine Biology Committee, and their Biological Station at Port Erin (Isle of Man). *Trans. Lpool biol. Soc.* **14,** 1–67.

Herdman, W. A. 1903. The new Biological Station at Port Erin (Isle of Man), being the seventeenth annual report of the Liverpool Marine Biology Committee. *Trans. Lpool biol. Soc.* **17,** 1–67.

Holme, N. A. 1949. The fauna of sand and mud banks near the mouth of the Exe estuary. *J. mar. biol. Ass. U.K.* **28,** 189–237.

Horsman, E. 1938. Additions to the marine fauna of Aberystwyth and district. *Aberyst. Stud.* **4,** 259–68.

Hubrecht, A. A. W. 1875. Untersuchungen über Nemertinen aus dem Golf von Neapel. *Niederl. Arch. Zool.* **2,** 99–135.

Hubrecht, A. A. W. 1879. The genera of European nemerteans critically revised, with description of several new species. *Notes Leyden Mus.* **1,** 193–232.

Hubrecht, A. A. W. 1880. New species of European nemerteans. *Notes Leyden Mus.* **2,** 93–8.

Humes, A. G. 1942. The morphology, taxonomy, and bionomics of the nemertean genus *Carcinonemertes. Illinois biol. Monogr.* **18,** 1–105.

Hunt, O. D. 1925. The food of the bottom fauna of the Plymouth fishing grounds. *J. mar. biol. Ass. U.K.* **13,** 560–99.

Hunt, O. D. 1948. Some freshwater and terrestrial worms from the Yealm basin. *Rep. Trans. Plymouth Instn*, **20,** 84.

Hylbom, R. 1957. Studies on palaeonemerteans of the Gullmar Fiord area (West coast of Sweden). *Ark. Zool.* **10** (2), 539–82.

Iwata, F. 1954. The fauna of Akkeshi Bay. XX. Nemertini in Hokkaido (revised report). *J. Fac. Sci. Hokkaido Univ., Zool.* **12** (6), 1–39.

Iwata, F. 1960. Studies on the comparative embryology of nemerteans with special reference to their interrelationships. *Publs. Akkeshi mar. biol. Stn.* No. 10, 1–51.

Jackson, L. H. 1935. Sense-organs in *Malacobdella. Nature, Lond.* **135,** 792.

Jameson, H. L. 1898. Notes on Irish worms: I. – The Irish nemertines, with a list of those contained in the Science and Art Museum, Dublin. *Proc. R. Ir. Acad.* **5** (3), 34–9.

Jameson, R. 1811. Catalogue of animals, of the class Vermes, found in the Firth of Forth, and other parts of Scotland. *Wernerian nat. Hist. Soc. Mem.* **1,** 556–65.

Jennings, J. B. 1960. Observations on the nutrition of the rhynchocoelan *Lineus ruber* (O. F. Müller). *Biol. Bull. mar. biol. Lab., Woods Hole*, **119,** 189–96.

Jennings, J. B. 1962. A histochemical study of digestion and digestive enzymes in the rhynchocoelan *Lineus ruber* (O. F. Müller). *Biol. Bull. mar. biol. Lab., Woods Hole*, **122,** 63–72.

Jennings, J. B. 1968. A new astomatous ciliate from the entocommensal rhynchocoelan *Malacobdella grossa* (O. F. Müller). *Arch. Protistenk.* **110,** 422–5.

Jennings, J. B. 1969. Ultrastructural observations on the phagocytic uptake of food materials by the ciliated cells of the rhynchocoelan intestine. *Biol. Bull. mar. biol. Lab., Woods Hole*, **137,** 476–85.

Jennings, J. B. & Gibson, R. 1968. The structure and life history of *Haplosporidium malacobdellae* sp. nov., a new sporozoan from the entocommensal rhynchocoelan *Malacobdella grossa* (O. F. Müller). *Arch. Protistenk.* **111,** 31–7.

Jennings, J. B. & Gibson, R. 1969. Observations on the nutrition of seven species of rhynchocoelan worms. *Biol. Bull. mar. biol. Lab., Woods Hole*, **136,** 405–33.

Johnston, G. 1827–28. Contributions to the British fauna. *Zool. J.* **3,** 486–91.

Johnston, G. 1828–29. Contributions to the British fauna. *Zool. J.* **4,** 52–7.

Johnston, G. 1833. Illustrations in British Zoology. *Loudon's Mag. nat. Hist.* **6,** 232–5.

Johnston, G. 1834. Illustrations in British Zoology. *Loudon's Mag. nat. Hist.* **7,** 584–8.

Johnston, G. 1837. Miscellanea Zoologica. II. A description of some planarian worms. *Mag. Zool. Bot.* **1,** 529–38.

Johnston, G. 1846. An index to the British annelides. *Ann. Mag. nat. Hist.* **16** (1), Supplement, 433–62.

Johnston, G. 1850. *An Introduction to Conchology; or, Elements of the Natural History of Molluscous Animals.* Van Voorst, London, 614 pp.

Johnston, G. 1865. *A Catalogue of the British Non-Parasitical Worms in the Collection of the British Museum.* Taylor & Francis, London, 365 pp.

Jones, A. M., Jones, Y. M. & James, J. L. 1979. The incidence of the nemertine *Malacobdella grossa* in the bivalve *Cerastoderma edule* in Shetland. *J. mar. biol. Ass. U.K.* **59,** 373–5.

Jones, H. D. 1978. Terrestrial planarians. *Lancs. Chesh. faun. Soc.* **73,** 19–20.

Jones, N. S. 1939. Some recent additions to the off-shore fauna of Port Erin. *Rep. mar. biol. Stn Port Erin*, **52,** 18–32.

Jones, N. S. 1951. The bottom fauna off the south of the Isle of Man. *J. anim. Ecol.* **20,** 132–44.

Joubin, L. 1890. Recherches sur les Turbellariés des cotes de France (Némertes). *Arch Zool. exp. gén.* **8** (2), 461–602.

*Joubin, L. 1894. Les Némertiens. *Faune Française*, ed. R. Blanchard & J. de Guerne, Société d'Éditions Scientifiques, Paris, 235 pp.

Judges, E. C. & Southward, A. J. 1953. A note on the populations in the sea-water storage tanks at the Port Erin laboratory. *Rep. mar. biol. Stn Port Erin*, **65,** 24–7.

Keferstein, W. 1862. Untersuchungen über die Nemertinen. *Z. wiss. Zool.* **12,** 51–90.

Khayrallah, N. & Jones, A. M. 1975. A survey of the benthos of the Tay Estuary. *Proc. R. Soc. Edinb. Ser. B*, **75,** 113–35.

King, H. 1939. Amphiporine, an active base from the marine worm *Amphiporus lactifloreus. J. chem. Soc.* Part 2, 1365–6.

King, L. A. L. 1911. Clyde marine fauna. Supplementary list. *Ann. Rep. mar. biol. Ass. West Scot., Millport*, 60–97.

Kingsley, C. 1859. *Glaucus; or, the Wonders of the Shore.* Macmillan, Cambridge, 230 pp.

Kirsteuer, E. 1963a. Zur Ökologie systematischer Einheiten bei Nemertinen. *Zool. Anz.* **170,** 343–54.

Kirsteuer, E. 1963b. Beitrag zur Kenntnis der Systematik und Anatomie der adriatischen Nemertinen (Genera *Tetrastemma*, *Oerstedia*, *Oerstediella*). *Zool. Jb. Anat.* **80,** 555–616.

Kirsteuer, E. 1967a. Marine, benthonic nemerteans: how to collect and preserve them. *Am. Mus. Novit.* No. 2290, 1–10.

Kirsteuer, E. 1967b. New marine nemerteans from Nossi Be, Madagascar. Results of the Austrian Indo-Westpacific Expedition 1959/60. *Zool. Anz.* **178,** 110–22.

Kirsteuer, E. 1974. Description of *Poseidonemertes caribensis* sp. n., and discussion of other taxa of Hoplonemertini Monostilifera with divided longitudinal musculature in the body wall. *Zool. Scr.* **3,** 153–66.

Kirsteuer, E. 1977. Remarks on taxonomy and geographic distribution of the genus *Ototyphlonemertes* Diesing (Nemertina, Monostilifera). *Mikrofauna Meeresboden*, **61,** 167–81.

Knight-Jones, E. W. & Nelson-Smith, A. 1977. Sublittoral transects in the Menai Straits and Milford Haven. *Biology of Benthic Organisms*, ed. B. F. Keegan, P. O. Ceidigh & P. J. S. Boaden, Pergamon Press, Oxford, pp. 379–89.

Koehler, R. 1885. Recherches sur la faune marine des îles Anglo-Normandes. *Bull. Soc. Sci. Nancy*, **7** (2), 51–120.

Kölliker, A. 1845. Über drei neue Gattungen von Würmern. *Verh. schweiz. naturf. Ges.* 86–98.
Lankester, E. R. 1866. Annelida and Turbellaria of Guernsey. *Ann. Mag. nat. Hist.* **17** (3), 388–90.
Laverack, M. S. & Blackler, M. 1974. *Fauna and Flora of St. Andrews Bay.* Scottish Academic Press, Edinburgh, 310 pp.
Leslie, G. & Herdman, W. A. 1881. *The Invertebrate Fauna of the Firth of Forth.* McFarlane & Erskine, Edinburgh, 106 pp.
Ling, E. A. 1969. The structure and function of the cephalic organ of a nemertine *Lineus ruber. Tissue Cell*, **1,** 503–24.
Ling, E. A. 1970. Further investigations on the structure and function of cephalic organs of a nemertine *Lineus ruber. Tissue Cell*, **2,** 569–88.
Ling, E. A. 1971. The proboscis apparatus of the nemertine *Lineus ruber. Phil. Trans. R. Soc., B*, **262,** 1–22.
Ling, E. A. & Willmer, E. N. 1973. The structure of the fore-gut of a nemertine, *Lineus ruber. Tissue Cell*, **5,** 381–92.
Little, C. & Boyden, C. R. 1976. Variations in the fauna of particulate shores in the Severn estuary. *Estuar. Coastal mar. Sci.* **4,** 545–54.
Logan, G. 1860. Report of the Committee on Marine Zoology; with a notice of the Sprat-fishing in the Firth of Forth. *Proc. R. phys. Soc.* **2,** 240–3.
Maclaren, N. H. W. 1901. On the blood vascular system of *Malacobdella grossa. Zool. Anz.* **24,** 126–9.
Marine Biological Association, 1904. Plymouth marine invertebrate fauna. *J. mar. biol. Ass. U.K.* **7,** 155–298.
Marine Biological Association, 1931. *Plymouth Marine Fauna*, 2nd ed., Plymouth, 371 pp.
Marine Biological Association, 1957. *Plymouth Marine Fauna*, 3rd ed., Plymouth, 457 pp.
Marion, A. F. 1873. Recherches sur les animaux inférieurs du Golfe de Marseille. Observations sur un nouveau némertien hermaphrodite (*Borlasia kefersteinii*). *Annls Sci. nat., Zool.* **17** (5), 6–23.
Markowski, S. 1962. Faunistic and ecological investigations in Cavendish Dock, Barrow-in-Furness. *J. anim. Ecol.* **31,** 43–51.
Massy, A. L. 1912. Report of a survey of trawling grounds on the coasts of Counties Down, Louth, Meath and Dublin. Part III. Invertebrate fauna. *Fish, Ireland Sci. Invest. 1911*, No. 1, 1–225.
McIntosh, W. C. 1867a. On the annelids of St. Andrews. *Rep. Brit. Ass.* 92–3.
McIntosh, W. C. 1867b. On the gregariniform parasite of *Borlasia. Trans. R. microsc. Soc.* **15,** 38–41.
McIntosh, W. C. 1868a. On the boring of certain annelids. *Ann. Mag. nat. Hist.* **2** (4), 276–95.
McIntosh, W. C. 1868b. Report on the annelids dredged off the Shetland Islands by Mr Gwyn Jeffreys, 1867–68. *Rep. 38th Meet. Brit. Ass., Norwich*, 336–40.
McIntosh, W. C. 1869. On the structure of the British nemerteans, and some new British annelids. *Trans. R. Soc. Edinb.* **25,** 305–433.
McIntosh, W. C. 1870. Note on the development of lost parts in the nemerteans. *J. Linn. Soc., Zool.* **10,** 251–4.
*McIntosh, W. C. 1873–74. A monograph of the British annelids. Part 1. The nemerteans. *Ray Soc. Publ.* 1–214.
McIntosh, W. C. 1875a. On *Valencinia armandi*, a new nemertean. *Trans. Linn. Soc. Lond.* **1** (2), 73–81.
McIntosh, W. C. 1875b. On *Amphiporus spectabilis*, de Quatrefages, and other nemerteans. *Q. Jl microsc. Sci.* **15,** 277–93.

McIntosh, W. C. 1875c. *The Marine Invertebrates and Fishes of St. Andrews.* A. & C. Black, Edinburgh, 186 pp.
McIntosh, W. C. 1876. On the central nervous system, the cephalic sacs, and other points in the anatomy of the Lineidae. *J. Anat. Physiol., Lond.* **10,** 231–52.
McIntosh, W. C. 1906a. Notes from the Gatty Marine Laboratory, St. Andrews. No. XXVII. Part 3. On bifid nemerteans (*Cerebratulus angulatus*, O.F.M. = *marginatus*, Renier?) from Aberdeen and Naples. *Ann. Mag. nat. Hist.* **17** (7), 74–8.
McIntosh, W. C. 1906b. Notes from the Gatty Marine Laboratory, St. Andrews. No. XXVII. Part 4. On *Amphiporus hastatus*, M'Intosh. *Ann. Mag. nat. Hist.* **17** (7), 78–81.
McIntosh, W. C. 1927. Additions to the marine fauna of St. Andrews since 1874. *Ann. Mag. nat. Hist.* **19** (9), 49–94.
Mellanby, H. 1951. *Animal Life in Fresh Water. A Guide to Fresh-Water Invertebrates.* Methuen, London, 4th ed. 296 pp.
Milne, H. & Dunnet, G. M. 1972. Standing crop, productivity and trophic relations of the fauna of the Ythan Estuary. *The Estuarine Environment*, ed. R. S. K. Barnes & J. Green, Applied Science, London, pp. 86–106.
Montagu, G. 1802. Unpublished notebook: Linnean Society Library, London.
Montagu, G. 1804. Description of several marine animals found on the south coast of Devonshire. *Trans. Linn. Soc. Lond.* **7,** 61–85.
Montagu, G. 1808. Vermes Mollusca. Unpublished manuscript, 294 pp: Linnean Society Library, London (illustrations to this manuscript, by Eliza Dorville, are in a separate volume).
Moore, H. B. 1937. Marine fauna of the Isle of Man. *Proc. Trans. Lpool biol. Soc.* **50,** 1–293.
Moore, J. & Gibson, R. 1981. The *Geonemertes* problem (Nemertea). *J. Zool., Lond.* **194,** 175–201.
Moore, P. G. 1973. The kelp fauna of northeast Britain. II. Multivariate classification: turbidity as an ecological factor. *J. exp. mar. Biol. Ecol.* **13,** 127–63.
Müller, O. F. 1774. *Vermium Terrestrium et Fluviatilium, seu Animalium Infusoriorum, Helminthicorum, et Testaceorum, non Marinorum, Succincta Historia.* Vol. 1, Part 2. Havniae et Lipsiae, Copenhagen, pp. 57–72.
Müller, O. F. 1776. *Zoologiae Danicae Prodromus, seu Animalium Daniae et Norvegiae Indigenarum Characteres, Nomina, et Synonyma Imprimis Popularium.* Hallageriis, Havniae, 282 pp.
Neill, P. 1807. Monthly memoranda in natural history. *Scots Mag. Edinb. Lit. Misc.* 804.
Newbigin, M. 1901. *Life by the Seashore.* Swan Sonnenschein, London, 344 pp.
Newell, G. E. 1954. The marine fauna of Whitstable. *Ann. Mag. nat. Hist.* **7** (12), 321–50.
Örsted, A. S. 1843. Forsog til en ny classification af Planarierne (Planariea Dugès) grundet paa mikroskopisk-anatomiske Undersogelser. *Naturhist. Tidsskr.* **4,** 519–81.
Oudemans, A. C. 1885. The circulatory and nephridial apparatus of the Nemertea. *Q. Jl microsc. Sci.* **25,** Supplement, 1–80.
Oxner, M. 1908. Sur de nouvelles espèces de Némertes de Roscoff et quelques remarques sur la coloration vitale. *Bull. Inst. océanogr. Monaco*, No. 127, 1–16.
Pantin, C. F. A. 1944. Terrestrial nemertines and planarians in Britain. *Nature, Lond.* **154,** 80.
Pantin, C. F. A. 1947. The nephridia of *Geonemertes dendyi. Q. Jl microsc. Sci.* **88** (3), 15–25.

Pantin, C. F. A. 1950. Locomotion in British terrestrial nemertines and planarians: with a discussion on the identity of *Rhynchodemus bilineatus* (Mecznikow) in Britain, and on the name *Fasciola terrestris* O. F. Müller. *Proc. Linn. Soc. Lond.* **162,** 23–37.

Pantin, C. F. A. 1961. *Geonemertes*: a study in island life. *Proc. Linn. Soc. Lond.* **172,** 137–52.

Pantin, C. F. A. 1969. The genus *Geonemertes. Bull. Br. Mus. nat. Hist., (Zool.),* **18,** 263–310.

Parfitt, E. 1867. A catalogue of the annelids of Devonshire, with notes and observations. *Rep. Trans. Devon. Ass. Adv. Sci.* **2,** 209–46.

Pennant, T. 1812. *British Zoology*. Vol. 4. Wilkie & Robinson, London, 379 pp.

Punnett, R. C. 1900. On some South Pacific nemertines collected by Dr Willey. *Zoological Results Based on Material from New Britain, New Guinea, Loyalty Islands and elsewhere, Collected during the Years 1895, 1896 and 1897 by Arthur Willey*, Part 5, Cambridge University Press, pp. 569–84.

Punnett, R. C. 1901a. On two new British nemerteans. *Q. Jl microsc. Sci.* **44,** 547–64.

Punnett, R. C. 1901b. *Lineus. Liverpool Marine Biology Committee Mem. typ. Br. mar. Pl. Anim.* No. 7, 1–37.

Purchon, R. D. 1948. Studies on the biology of the Bristol Channel. XVII. The littoral and sublittoral fauna of the northern shores, near Cardiff. *Proc. Bristol Nat. Soc.* **27,** 285–310.

Purchon, R. D. 1956. Studies on the biology of the Bristol Channel. XVIII. The marine fauna at five stations on the northern shores of the Bristol Channel and Severn Estuary. *Proc. Bristol Nat. Soc.* **29,** 213–26.

Quatrefages, A. De, 1846. Études sur les types inférieurs de l'embranchement des annelés. Mémoire sur la famille des Némertiens (Nemertea). *Annls Sci. nat.* **6** (3), 173–303.

Rathke, H. 1843. Beiträge zur Fauna Norwegens. *Nova Acta Acad. Caesar. Leop. Carol.* **12,** 1–264.

Rathke, J. 1799. Jagttagelser henhörende til Indvoldeormenes og Blöddyrenes Naturhistorie. *Skr. nat. Selsk. Kbh.* **5,** 61–153.

Reisinger, E. 1926. Nemertini. Schnurwürmer. *Biologie Tiere Dtl.* **17,** 7.1–7.24.

Remane, A. 1958. Ökologie des Brackwassers. *Die Binnengewässer. Einzeldarstellungen aus der Limnologie und ihren Nachbargebieten.* Vol. 22. Die Biologie des Brackwassers, ed. A. Thienemann, E. Schweizerbart'sche Verlag, Stuttgart, pp. 1–216.

Renier, S. A. 1804. Prospetto della Classe dei Vermi, nominati e ordinati secondo il Sistema di Bosc. An uncompleted work issued in G. Meneghini's introduction to *Osservazione postume di Zoologica Adriatica del Professore Stefano Andrea Renier*, Venice, 1847, pp. 57–66.

Renouf, L. P. W. 1931. Preliminary work of a new biological station (Lough Ine, Co. Cork, I.F.S.). *J. Ecol.* **19,** 410–38.

Riches, T. H. 1893. A list of the nemertines of Plymouth Sound. *J. mar. biol. Ass. U.K.* **3,** 1–29.

Schultze, M. S. 1851. *Beiträge zur Naturgeschichte der Turbellarien.* Koch, Greifswald, 78 pp.

Scott, T. 1893. Notes on Forth Annelida. *Ann. Scot. nat. Hist.* 185–6.

Scott, T. 1894. On the occurrence of *Cerebratulus angulatus* (O. F. Müller) in the Firth of Forth. *Ann. Scot. nat. Hist.* 118–19.

Sheldon, L. 1896. Nemertinea. *The Cambridge Natural History*, ed. S. F. Harmer & A. E. Shipley, vol. 2, Macmillan, New York, pp. 97–120.

Sheppard, E. M. 1935. On *Paradrepanophorus crassus* (Quatr.), a nemertean worm new to the British fauna. *Ann. Mag. nat. Hist.* **15** (10), 232–6.

Slinger, I. 1975. Biochemical observations on the arylamidase enzymes of four species of nemertean worms. *Comp. Biochem. Physiol.* **50C,** 1–4.

Slinger, I. & Gibson, R. 1974. Biochemical observations on the phosphatase enzymes of five species of nemertean worms. *Comp. Biochem. Physiol.*, **47B,** 279–88.

Slinger, I. & Gibson, R. 1975. Biochemical studies on the esterase enzymes of four species of nemertean worms. *J. exp. mar. Biol. Ecol.* **17,** 95–102.

Smith, J. E. 1935. The early development of the nemertean *Cephalothrix rufifrons. Q. Jl Microsc. Sci.* **77,** 335–81.

Southern, R. 1908a. Nemertinea. *Handbook to the City of Dublin and the Surrounding District*, ed. G. A. J. Cole & R. L. Praeger, Dublin University Press, 441 pp.

Southern, R. 1908b. Occurrence of a fresh-water nemertine in Ireland. *Nature, Lond.* **79,** 8.

Southern, R. 1911. Some new Irish worms. *Ir. Nat.* **20,** 5–9.

Southern, R. 1913. Nemertinea. *Proc. R. Ir. Acad.* **31,** No. 55, 1–20.

Southward, A. J. 1953. The fauna of some sandy and muddy shores in the south of the Isle of Man. *Proc. Trans. Lpool biol. Soc.* **59,** 51–71.

Sowerby, J. 1806. *The British Miscellany: or Coloured Figures of New, Rare, or Little Known Animal Subjects; many not before Ascertained to be Inhabitants of the British Isles.* Taylor, London, 136 pp.

Stephenson, J. 1911. The nemertines of Millport and its vicinity. *Trans. R. Soc. Edinb.* **48,** 1–29.

Stiasny-Wijnhoff, G. 1926. The Nemertea Polystilifera of Naples. *Pubbl. Staz. zool. Napoli,* **7,** 119–68.

Stiasny-Wijnhoff, G. 1930. Die Gattung *Oerstedia. Zoöl. Meded., Leiden,* **13,** 226–40.

Stiasny-Wijnhoff, G. 1934. Some remarks on North Atlantic non-pelagic Polystylifera. *Q. Jl microsc. Sci.* **77,** 167–90.

Stiasny-Wijnhoff, G. 1936. Die Polystilifera der Siboga-Expedition. *Siboga Exped.* **22,** 1–214.

Stiasny-Wijnhoff, G. 1938. Das Genus *Prostoma* Duges, eine Gattung von Süsswasser-Nemertinen. *Archs néerl. Zool.* **3,** Supplement, 219–30.

Stimpson, W. 1857. Prodromus descriptionis animalium evertebratorum quae in Expeditione ad Oceanum Pacificum Septemtrionalem a Republica Federata missa, Cadwaladaro Ringgold et Johanne Rodgers Ducibus, observavit et descripsit. Pars II. Turbellarieorum Nemertineorum. *Proc. Acad. nat. Sci. Philad.*, 159–65.

Stopford, S. C. 1951. An ecological survey of the Cheshire foreshore of the Dee estuary. *J. anim. Ecol.* **20,** 103–22.

Sumner, J. C. 1894. Description of a new species of nemertine. *Ann. Mag. nat. Hist.* **14** (6), 114.

Tanner, F. L. 1908. Report of section for Marine Zoology. *Rep. Soc. nat. Sci. Guernsey*, 282–5.

Templeton, R. 1836. A catalogue of the species of annulose animals, and of rayed ones, found in Ireland, as selected from the papers of the late J. Templeton, Esq., of Cranmore, with localities, descriptions, and illustrations. *Mag. nat. Hist.* **9,** 233–40.

Thompson, T. E., Smith, S. J., Jenkins, M., Benson-Evans, K., Fisk, D., Morgan, G., Delhanty, J. E. & Wade, A. E. 1966. Contributions to the biology of the Inner Farne. *Trans. nat. Hist. Soc. Northumb.* **15,** 197–225.

Thompson, W. 1838. Notes upon the natural history of a portion of the south west of Scotland. *Charlesworth's Mag. nat. Hist.* **2,** 18–21.

Thompson, W. 1841. Additions to the fauna of Ireland. *Ann. Mag. nat. Hist.* **7** (1), 477–82.

Thompson, W. 1843. Report on the fauna of Ireland: Div. Invertebrata. *Rep. Br. Ass.* 245–91.

Thompson, W. 1845. Additions to the fauna of Ireland, including descriptions of some apparently new species of Invertebrata. *Ann. Mag. nat. Hist.* **15** (1), 308–22.

Thompson, W. 1846. Additions to the fauna of Ireland, including a few species unrecorded in that of Britain; – with the description of an apparently new *Glossiphonia. Ann. Mag. nat. Hist.* **18** (1), 383–97.

Thompson, W. 1856. *The Natural History of Ireland.* Vol. IV. Mammalia, Reptiles, and Fishes, also Invertebrata. Bohn, London, 516 pp.

Todd, R. A. 1905. Report on the food of fishes collected during 1903. *First Report on Fishery and Hydrographical Investigations in the North Sea and Adjacent Waters (Southern Area)*, H.M.S.O., London, pp. 227–87.

Turton, W. 1807. *British Fauna, Containing a Compendium of the Zoology of the British Islands: Arranged According to the Linnean System.* Vol. 1. Including the Classes Mammalia, Birds, Amphibia, Fishes and Worms. Evans, Swansea, 230 pp.

Vaillant, L. 1890. Lombriciniens, Hirudiniens, Bdellomorphes, Térétulariens et Planariens. *Hist. nat. Ann. mar. d'eau douce, Paris*, **3,** Part 2, 341–768.

Vanstone, J. H. & Beaumont, W. I. 1894. Report upon the nemertines found in the neighbourhood of Port Erin, Isle of Man. *Trans. Lpool biol. Soc.* **8,** 135–9.

Vanstone, J. H. & Beaumont, W. I. 1895. Report upon the nemertines found in the neighbourhood of Port Erin, Isle of Man. *Lpool mar. biol. Comm. Rep.* **4,** 216–20.

Varndell, I. M. 1980a. Oxygen consumption and the occurrence of a haemoprotein within the central nervous tissue of the hoplonemertean *Amphiporus lactifloreus* (Johnston) (Nemertea: Enopla: Monostilifera). *J. exp. mar. Biol. Ecol.* **45,** 157–72.

Varndell, I. M. 1980b. The occurrence, life history and effects of *Haplosporidium malacobdellae* Jennings and Gibson in a new host, the monostiliferous hoplonemertean *Amphiporus lactifloreus* (Johnston). *Arch. Protistenk.* **123,** 382–90.

Varndell, I. M. 1981a. A histochemical study of *Haplosporidium malacobdellae*, a balanosporidan endoparasite of the hoplonemertean *Amphiporus lactifloreus. Z. Parasitenkd.* **65,** 143–51.

Varndell, I. M. 1981b. Catechol oxidase activity associated with sporogenesis in *Haplosporidium malacobdellae*, a balanosporidan endoparasite of the hoplonemertean *Amphiporus lactifloreus. Z. Parasitenkd.* **65,** 153–62.

Verrill, A. E. 1892. The marine nemerteans of New England and adjacent waters. *Trans. Conn. Acad. Arts Sci.* **8,** 382–456.

Walton, C. L. 1913. The shore fauna of Cardigan Bay. *J. mar. biol. Ass. U.K.* **10,** 102–13.

Waterston, A. R. & Quick, H. E. 1937. *Geonemertes dendyi* Dakin, a land nemertean, in Wales. *Proc. R. Soc. Edinb.* **57,** 379–84.

Wheeler, J. F. G. 1934. Nemerteans from the South Atlantic and southern oceans. *'Discovery' Rep.* **9,** 215–94.

Whitfield, P. J. 1972. The ultrastructure of the spermatozoon of the hoplonemertine, *Emplectonema neesii. Z. Zellforsch.* **128,** 303–16.

Wijnhoff, G. 1912. List of nemerteans collected in the neighbourhood of Plymouth from May–September, 1910. *J. mar. biol. Ass. U.K.* **9,** 407–34.

Wijnhoff, G. 1913. Die Gattung *Cephalothrix* und ihre Bedeutung für die Systematik der Nemertinen. II. Systematischer Teil. *Zool. Jb. (Syst.)* **34,** 291–320.

Williams, R. B. 1972. Notes on the history and invertebrate fauna of a poikilohaline lagoon in Norfolk. *J. mar. biol. Ass. U.K.* **52,** 945–63.

Williams, T. 1852. Report on the British Annelida. *Rep. 21st Meet. Br. Ass., Ipswich 1851*, 159–272.
Williams, T. 1858. Researches on the structure and homology of the reproductive organs of the annelids. *Phil. Trans. R. Soc.* **148,** 93–144.
Wilson, C. B. 1900. The habits and early development of *Cerebratulus lacteus* (Verrill). *Q. Jl microsc. Sci.* **43,** 97–198.
Withers, R. G. & Thorp, C. H. 1977. Studies on the shallow, sublittoral epibenthos of Langstone Harbour, Hampshire, using settlement panels. In *Biology of Benthic Organisms*, ed. B. F. Keegan, P. O. Ceidigh & P. J. S. Boaden, Pergamon Press, Oxford, pp. 595–604.

Index to families, genera and species

Family names are listed in capital letters, valid names for genera and species in italic type, and synonyms and dubious species in roman type. Page citations in roman type refer to the text, those in italics to illustrations, and those in heavy type to distribution maps.